39th Conference on Recent Developments on High Energy Physics and Cosmology (HEP 2022)

AA004016

Journal of Physics: Conference Series Volume 2375

Thessaloniki, Greece
15-18 June 2022

ISBN: 978-1-7138-6482-0
ISSN: 1742-6588

Printed from e-media with permission by:

Curran Associates, Inc.
57 Morehouse Lane
Red Hook, NY 12571

Some format issues inherent in the e-media version may also appear in this print version.

Institute of Physics
Dirac House, Temple Back
Bristol BS1 6BE UK

Phone: 44 1 17 929 7481
Fax: 44 1 17 920 0979

techtracking@iop.org

Additional copies of this publication are available from:

Curran Associates, Inc.
57 Morehouse Lane
Red Hook, NY 12571 USA
Phone: 845-758-0400
Fax: 845-758-2633
Email: curran@proceedings.com
Web: www.proceedings.com

TABLE OF CONTENTS

Preface

Peer Review Statement

Precision Measurement of the W Boson Mass Using the Full CDF Run II Data Set 1
Costas Vellidis

First Search for Magnetic Monopoles Through the Schwinger Mechanism 11
Vasiliki A Mitsou

The CMS Trigger System 21
Milos Dordevic

Performance Studies of Micromegas Electronics in a High Radiation Environment at the CERN Gamma Irradiation Facility (GIF++) 31
Foteini Kolitsi, Daniel Bita, Ioannis Mesolongitis, Chara Kitsaki, Theodoros Alexopoulos, Efstathios Kyriakis-Bitzaros, Katerina Zachariadou

Segment Linking: A Highly Parallelizable Track Reconstruction Algorithm for HL-LHC 42
P Chang, P Elmer, Y Gu, V Krutelyov, G Niendorf, M Reid, B V Sathia Narayanan, M Tadel, E Vourliotis, B Wang, P Wittich, A Yagil

The Innovative FEL Design by the CompactLight Collaboration 51
N Gazis, A Bignami, E Trachanas, T Alexopoulos, I Telali, T Apostolopoulos, K Pramatari, A Karagiannaki, D Kotsopoulos, E Gazis, D V Bandekas

Development of a Muon Tomography Application with Micromegas Detectors 62
D Amperiadou, C Petridou, D Sampsonidis, S Tzamarias, K Kordas, C Lampoudis, A Leisos, A Tsirigotis, Ch Tsiafis

Air Shower Radio Signal Electric Field Orientation as Measured with the Astroneu Cosmic Ray Telescope. 72
S Nonis, A Leisos, A Tsirigotis, G Bourlis, K Papageorgiou, I Gkialas, I Manthos, S Tzamarias

Measurement of the Higgs Quartic Coupling c_2v from di-Higgs Vector Boson Fusion in the $bb\tau^+\tau^-$ Channel 82
Christos Paraskevopoulos

The New Small Wheel High Voltage Infrastructure 92
I. Drivas-Koulouris, Efstathios Karentzos

The Detector Control System of the New Small Wheel Electronics for the ATLAS Experiment 100
Polyneikis Tzanis

Formulating E- & T-Model Inflation in Supergravity 110
C Pallis

Sterile Neutrino from D-Brane Models 120
Christos Kokorelis

Author Index

Preface

HEP 2022 - 39th Conference on Recent Developments in High Energy Physics and Cosmology was successfully held online between the 15th to the 18th of June 2022, in Thessaloniki, Greece. The conference was jointly organized and sponsored by the Hellenic Society for the Study of High Energy Physics, the Aristotle University of Thessaloniki and the Research Committee of the Aristotle University.
The conference invited scholars and experts in the fields of High Energy Physics, involved both in experimental and theoretical research, from various universities and institutions. The conference focuses on the more recent advancements in topics such as 'LHC experimental results', 'Astroparticle Physics', 'Cosmology', 'String Theory', 'Dark Matter' and others. The main goal of the conference is to present the related research work of the Greek scientific community, to promote the cooperation of experts and scholars in the field, to share the research achievements both on a national and international level. All accepted speakers were eligible to contribute to the Conference Proceedings on a voluntary basis.

This collection of Proceedings compiles oral and paper presentations submitted by the authors and scrutinized by the Editorial Committee. The HEP2022 Conference venue took place at Aristotle University's Research Dissemination Center (KEDEA Building), at Aristotle University's campus between the Student Club and the University Gymnasium.

The opening ceremony was held in the Amphitheatre I of KEDEA Building and a welcome speech was delivered by the chair of the Conference, Professor Kostas Kordas. Each author delivered and oral talk of 35 minutes accompanied with the slides material, along with 5 minutes extra time, allocated for questions and discussion. A dedicated convener chaired each plenary session.
The organizing and editorial committee of the HEP2022 hopes you enjoy the reading of this volume of the JPCS. Additionally, we want to thank all authors and participants for providing their valuable contributions for this proceeding as well as the reviewers for their constructive recommendations and feedback aiding to improve the presented articles.

Organizing Committee

K.Kordas (Chair), Aristotle University of Thessaloniki

A.Kyriakis, National Centre for Scientific Research Demokritos
K. Papageorgiou, University of Agean
A. Petkou, Aristotle University of Thessaloniki
Ch.Petridou, Aristotle University of Thessaloniki
D.Sampsonidis, Aristotle University of Thessaloniki, Greece
V. Spanos, National and Kapodistrian University of Athens
S.E.Tzamarias, Aristotle University of Thessaloniki
N.Vlachos, Aristotle University of Thessaloniki
K.Zachariadou, University of West Attica

Peer Review Statement

All papers published in this volume have been reviewed through processes administered by the Editors. Reviews were conducted by expert referees to the professional and scientific standards expected of a proceedings journal published by IOP Publishing Publishing.

- **Type of peer review:** Single Anonymous
- **Conference submission management system:** Morressier
- **Number of submissions received:** 13
- **Number of submissions sent for review:** 13
- **Number of submissions accepted:** 13
- **Acceptance Rate (Submissions Accepted / Submissions Received × 100):** 100
- **Average number of reviews per paper:** 1
- **Total number of reviewers involved:** 9
- **Contact person for queries:**
  Name: Dimos Sampsonidis
  Email: sampson@physics.auth.gr
  Affiliation: Aristotle University of Thessaloniki, Greece

Precision measurement of the W boson mass using the full CDF Run II data set

Costas Vellidis, for the CDF Collaboration

National and Kapodistrian University of Athens, Physics Department, University Campus, Zografou 15784, Greece

E-mail: konstantinos.vellidis@cern.ch

Abstract. The mass of the W boson, a mediator of the weak force between elementary particles, is tightly constrained by the symmetries of the standard model of particle physics. After the observation of the Higgs boson, the last missing component of the model, the measurement of the W boson mass provides a stringent test of the model. A measurement of the W boson mass is presented, which used data corresponding to 8.8 fb^{-1} of integrated luminosity, collected in proton-antiproton collisions at a 1.96 TeV center-of-mass energy with the CDF II detector at the Fermilab Tevatron collider. A sample of approximately 4 million W boson candidates was used to obtain a W boson mass of $80433.5 \pm 6.4_{\mathrm{stat}} \pm 6.9_{\mathrm{syst}} = 80433.5 \pm 9.4$ MeV, the precision of which exceeds that of all previous measurements combined. This measurement is in significant tension with the standard model expectation.

1. Introduction

The W boson, together with the Z boson, is the quantum carrier of the weak nuclear force, which is responsible for radioactive phenomena, such as the nuclear beta decay. The standard model (SM) of particle physics describes jointly the weak and electromagnetic forces in terms of four, originally massless, vector (spin-1) fields based on the so-called electroweak gauge symmetry. This symmetry is spontaneously "broken" by the scalar (spin-0) Higgs field, also having four degrees of freedom, three of which are absorbed by three vector fields, thus making $W^{\pm}$ and Z massive and leaving one vector field, the photon, massless. The discovery of the particle manifestation of the remaining fourth degree of freedom of the Higgs field, the Higgs boson H, at the LHC in 2012 confirmed this picture of fundamental forces and tied together the W, Z and H bosons.

The electroweak gauge sector of the SM is described by three free parameters, namely the gauge couplings of the W and Z bosons and the vacuum expectation value of the Higgs field. These fundamental parameters are constrained by three precisely measured observables: the fine structure constant α, determined from measurements of electromagnetic phenomena, the Fermi constant G_F, determined from the muon lifetime, and the (pole) mass of the Z boson, determined from the data taken by the LEP experiments during the LEP energy scan. These three observables constrain, in turn, other electroweak observables, such as the W boson mass M_W [1]:

$$\mathrm{M_W} = \frac{\mathrm{M_Z}}{\sqrt{2}}\sqrt{1+\sqrt{1-\frac{\sqrt{8}\pi\alpha}{\mathrm{G_F M_Z^2}}(1+\Delta\mathrm{r})}} \tag{1}$$

A precise measurement of $\mathrm{M_W}$ can be compared with the SM expectation from Eq. 1, thus providing a very sensitive probe of the internal consistency of the SM.

The term $\Delta\mathrm{r}$ in Eq. 1 originates from virtual processes affecting $\mathrm{M_W}$, such as the W boson splitting into a top-bottom quark pair annihilating back to W, or the emission and re-absorption of a H, Z or γ boson by the W boson. The effects of such processes on $\mathrm{M_W}$ are collectively called "loop corrections" and they are systematically calculated as perturbations to the free W propagation. The most important one-loop corrections introduce a quadratic dependence of $\mathrm{M_W}$ on the top quark mass $\mathrm{m_t}$ and a logarithmic dependence on the Higgs boson mass $\mathrm{M_H}$, both of which are precisely measured [2]:

$$\Delta\mathrm{r} \simeq \frac{\alpha}{\pi}\frac{\mathrm{M_Z^2}}{\mathrm{M_Z^2}-\mathrm{M_W^2}}\left(\frac{11}{24}\ln\frac{\mathrm{M_H}}{\mathrm{M_Z}} - \frac{3}{16\pi}\frac{\mathrm{m_t^2}}{\mathrm{M_Z^2}-\mathrm{M_W^2}}\right) + \Delta\alpha \tag{2}$$

Somewhat more involved is the photon loop contribution $\Delta\alpha$, which brings in a non-perturbative correction for virtual hadron processes. Nonetheless, the SM expectation [3] $\mathrm{M_W} = 80\,357 \pm 4_{\mathrm{inputs}} \pm 4_{\mathrm{theory}} = 80\,357 \pm 6$ MeV is very accurate. Global fits of precision electroweak observables verify the SM expectation down to the MeV level.

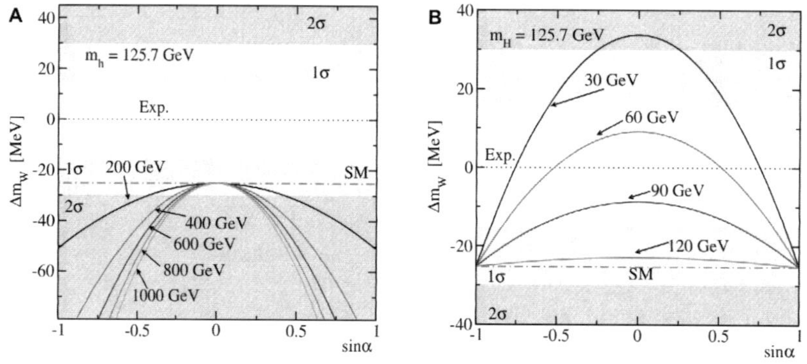

Figure 1. Variation of the W boson mass with the mixing of the SM Higgs boson and a hypothetical new scalar particle having no SM charges. The dashed-dotted line shows the SM expectation and the solid lines show the expectations for various hypotheses of the new particle's mass. The dotted line shows the experimental result with the 1-σ and 2-σ uncertainty bands in yellow and cyan, respectively. (**A**) Scenario 1, where the light scalar is the SM Higgs boson and the heavy scalar is the new particle. (**B**) Scenario 2, where the heavy scalar is the SM Higgs boson and the light scalar is the new particle.

The W boson mass is also a sensitive probe for physics beyond the SM (BSM). Similar to the SM case, $\mathrm{M_W}$ can be affected by virtual processes involving particles proposed by various BSM theories. Examples are theories attempting to provide a deeper explanation of the Higgs sector of the SM, such as supersymmtery, compositeness, theories with new strong interactions, and theories with extended Higgs sector; or theories of dark matter particles; or theories involving an extended gauge sector. An illustrative example is the inclusion, in addition to the Higgs boson, of a new scalar particle with no SM charges, which mixes with the SM Higgs boson [4]. The sensitivity of $\mathrm{M_W}$ to the effect of such a new particle is shown in Fig. 1, displaying the change of M_W as a function of the sine of the mixing angle under various hypotheses for the mass of the new particle.

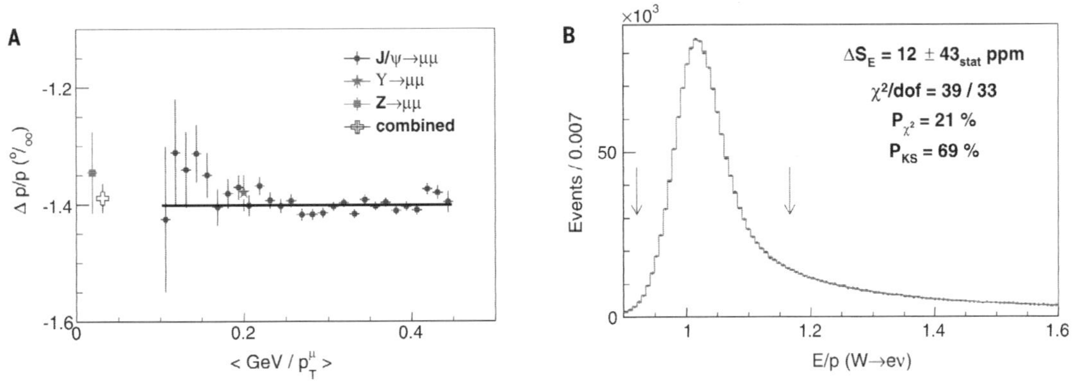

Figure 2. Lepton momentum calibration. (**A**) Tracker momentum scale correction $\Delta p/p$ as a function of the average $1/p_T^{\mu}$ per bin in $J/\psi \rightarrow \mu\mu$ data (blue circles), $\Upsilon \rightarrow \mu\mu$ data (one bin, red star), $Z \rightarrow \mu\mu$ data (one bin, green square), and combined (purple cross). (**B**) The fit of the simulated template (red histogram) of the electron energy-to-momentum ratio E/p to the data (blue points). The arrows mark the fitting window. The legend shows the χ^2/dof, the χ^2 probability P_{χ^2}, and the Kolmogorov-Smirnov probability P_{KS} of the fit.

This article describes [5] a precise measurement of M_W using the full Run II dataset of the CDF experiment at the Fermilab Tevatron proton-antiproton collider operating during 1992-95 (Run I) at 1.8 TeV and 2001-2011 (Run II) at 1.96 TeV centre-of-mass energy. The dataset was collected with the Collider Detector at Fermilab (CDF), which was a multi-purpose detector, nearly hermetic and cylindrically symmetric around the axis of the colliding beams, consisting from inside outwards of a silicon vertex detector; a drift wire chamber called Central Outer Tracker (COT); a superconducting toroid producing a magnetic field of 1.4 T; a sampling calorimeter with an electromagnetic (innermost) and an hadronic (outermost) compartment; and proportional wire chambers for muon detection. The lepton acceptance in pseudorapidity $\eta = \ln\cot(\theta/2)$ (in a spherical coordinate system $\{r, \theta, \phi\}$) extended out to $|\eta| < 1.5$ for muons and $|\eta| < 2.0$ for electrons and the jet acceptance out to $|\eta| < 2.8$, with b-tagging capability out to $|\eta| < 1.4$.

2. Analysis strategy

At the Tevatron, the W boson was produced ~80% of the time via quark-antiquark annihilation. It can decay either hadronically, into a pair of quarks producing a pair of jets, or leptonically, into a charged lepton plus a neutrino. The hadronic final states, including the one involving the decay into a tau lepton subsequently decaying into hadrons, have very poor resolution to allow for a precise measurement of M_W. The leptonic final states involving an electron or muon, on the other hand, can only be partially reconstructed, due to the undetected neutrino, but the lepton detection has a very good resolution allowing for a precise measurement of M_W. The measurement is using three kinematic variables sensitive to M_W: the lepton transverse momentum $\vec{p}_T^{\ell}$, the neutrino transverse momentum $\vec{p}_T^{\nu} = -\sum \vec{p}_T$, where the sum extends over all observed final state objects, and the W boson transverse mass $m_T = \sqrt{2p_T^{\ell}p_T^{\nu}[1-\cos(\phi_\ell - \phi_\nu)]}$. The lepton transverse momentum $p_T^{\ell} = p_\ell \sin\theta_\ell$ carries most of the information on M_W and can be calibrated very precisely, with a relative uncertainty of 0.004%. The hadronic activity from

Journal of Physics: Conference Series **2375** (2022) 012001 doi:10.1088/1742-6596/2375/1/012001

initial state gluon radiation and proton breakup is of the order of 10 GeV and is measured in the hadronic calorimeter, calibrated to $\sim 0.2\%$. Non-W backgrounds are small and partly measured from data. Templates of $p_T{}^{\ell}$, $p_T{}^{\nu}$, and m_T are simulated as functions of M_W using accurate models of the structure of the colliding protons, of the W boson production mechanism, and of the W boson decay mechanism. These templates are then fitted to the data to measure M_W.

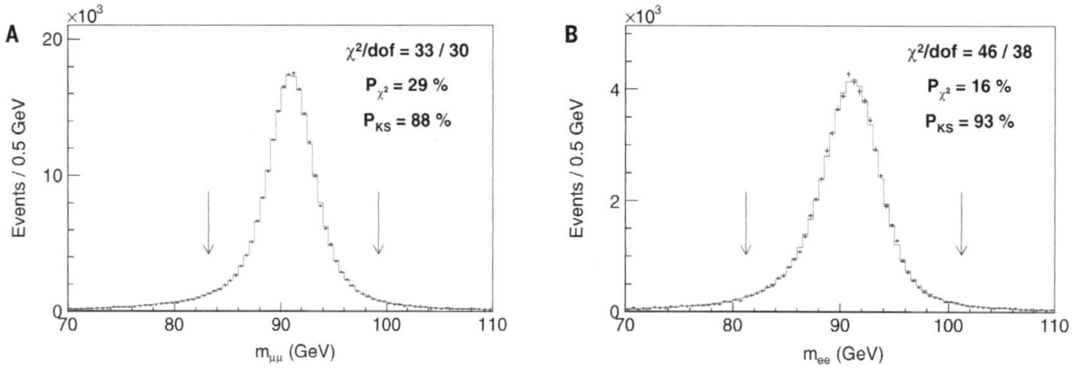

Figure 3. Fits of the Z boson mass simulated templates (red histograms) to the data (blue points). The arrows mark the fitting windows. The legends show the χ^2/dof, the χ^2 probability P_{χ^2}, and the Kolmogorov-Smirnov probability P_{KS} of the fits. (**A**) Fit to the $Z \to \mu\mu$ data after the tracker momentum calibration using low mass dimuon resonances. (**B**) Fit to the $Z \to ee$ data after the calorimeter energy calibration using the electron energy-to-momentum ratio E/p.

In order to maximise the M_W information content and minimise the backgrounds, events were selected with high-p_T central ($|\eta| < 1$) leptons and small hadronic recoil activity using inclusive lepton triggers that required a loose track, a muon segment or a calorimeter cluster, and lepton $p_T > 18$ GeV. The kinematic efficiency of the triggers was $\sim$100% for the offline event selection. The offline lepton selection required an electron cluster with transverse energy $E_T = E \sin\theta > 30$ GeV or a muon track with $p_T > 30$ GeV, plus loose lepton identification criteria to minimise bias. The W boson event selection required one selected lepton, hadronic recoil $u < 15$ GeV ($\vec{u} = \sum \vec{p}_T$ where the sum extends over all hadronic objects) and missing transverse energy (i.e. $p_T{}^{\nu}$) MET > 30 GeV. Z boson events were also selected as a calibration and control sample. The Z boson event selection required two selected leptons. With these criteria, 1 811 700 $W \to e\nu_e$, 66 180 $Z \to ee$, 2 424 486 $W \to \mu\nu_\mu$, and 238 534 $Z \to \mu\mu$ events were selected. These samples, corresponding to 8.8 fb^{-1} of integrated luminosity for both the electron and muon channels, were fairly clean, with mis-identification backgrounds about 0.5% of the signal. The running conditions were identical for both channels, thus guaranteeing cross-calibration.

The energy scale measurements determine the precision of the M_W measurement, therefore a detailed program of lepton energy scales was deployed. First, the 30 240 wires (arranged in 2 520 cells) of the COT were aligned cell-by-cell using a clean sample of about 480 000 cosmic rays [6]. The cell relative alignment was thus reduced from $\sim$50 μm down to $\sim$1 μm, improving significantly the track momentum resolution. Next, the COT momentum scale and non-linearity were constrained using $J/\psi \to \mu\mu$ and $\Upsilon \to \mu\mu$ mass fits, and the calibration was confirmed using a $Z \to \mu\mu$ mass fit. Then, the COT momentum scale was transferred to the electromagnetic (EM) calorimeter using a fit to the peak of the energy-to-momentum (E/p) ratio spectrum around $E/p \sim 1$, and the calorimeter energy scale was confirmed using a $Z \to ee$ mass fit. Finally, the COT and EM calorimeter resolutions were determined and the hadronic recoil was modelled, to reconstruct accurately the W boson kinematics and build the templates for the M_W fits. All M_W and M_Z fit results were blinded with a random offset in the range

Table 1. Estimated non-W background fractions in the data.

Source	Fraction in the muon channel (%)	Fraction in the electron channel (%)
$Z\gamma^* \to \mu\mu$	7.37±0.10	0.134±0.003
$W \to \tau\nu$	0.880±0.004	0.94±0.01
Hadronic jets	0.01±0.04	0.34±0.08
Decays in flight	0.20±0.14	—
Cosmic rays	0.01±0.01	—
Total	8.47±0.18	1.41±0.08

[−50,50] MeV. The offset was removed after the analysis was frozen. This technique allows to study all aspects of the data while keeping the mass fit results unknown within ±50 MeV.

3. Simulation

The NNPDF3.1 set [7] at next-to-next-to-leading order (NNLO) in α_s was used for the parton density functions (PDF) in the proton. These densities are generated by using gaussian-sampled "replicas" of data as inputs to machine-learning (ML) algorithms. For the uncertainty estimation, 25 eigenvector PDF sets were derived from 1000 replicas. The variation δM_W in the W boson mass from each eigenvector PDF was computed and an uncertainty of 3.9 MeV on M_W was estimated from the r.m.s. of the M_W values obtained from the 25 PDF eigenvectors. The central M_W value obtained with the NNPDF3.1 set at NNLO and with other NNLO sets (CT18, MMHT2014) agreed within 2.1 MeV. The central M_W value obtained with the NNPDF3.1 set at NLO and with other NLO sets (ABMP16, CJ15, MMHT2014) agreed within 3 MeV. The uncertainty on M_W from missing higher-order QCD effects was estimated to be 0.4 MeV by varying the factorisation and renormalisation scales, and by comparing the results obtained with two event generators using different models of soft gluon radiation (ResBos and MadGraph aMC@NLO + Pythia).

The vector boson production model must account for the hadronic activity in the initial state. The perturbative expansion of $d\sigma/dp_T^V$ (V = W or Z) has terms proportional to $\alpha_s^n \ln^{2n}(p_T^2/M_V^2)$. The series thus diverges as $p_T^V \to 0$. It is therefore needed to include corrections at all orders by resumming the series. There are two resummation methods: (a) the analytical method, where a formal resummation of the perturbation series is performed at low p_T^V and the result is matched to a fixed-order matrix element calculation at high p_T^V; and (b) the numerical method using parton-shower Monte Carlo (MC) simulation. The vector boson decay model must account for spin correlations among the vector boson and the final state leptons, and for EM radiation from all electrically charged particles in the process.

The model used by CDF for the W or Z boson production and decay is provided by the ResBos analytical resummation program [8], whereas the model used for EM radiation is generated with the Photos program [9] and validated by comparing with the Horace program [10]. The ResBos program calculates the 5-fold cross section $d\sigma/(dp_T^V dy_V dM_V d\cos\theta_\ell d\phi_\ell)$, fully differential in the vector boson kinematics, p_T^V, rapidity $y_V = (1/2)\ln[(E_V + p\cos\theta_V)/(E_V - p\cos\theta_V)]$, and mass; and in the angles of emission of the charged lepton. The calculation accuracy is NLO for the matrix element and NNLL (next-to-next-to-leading log) for the resummation. The Photos program accounts for multi-photon emission only from the final-state charged lepton(s). Validation of Photos by comparing the results with those from the Horace program, which

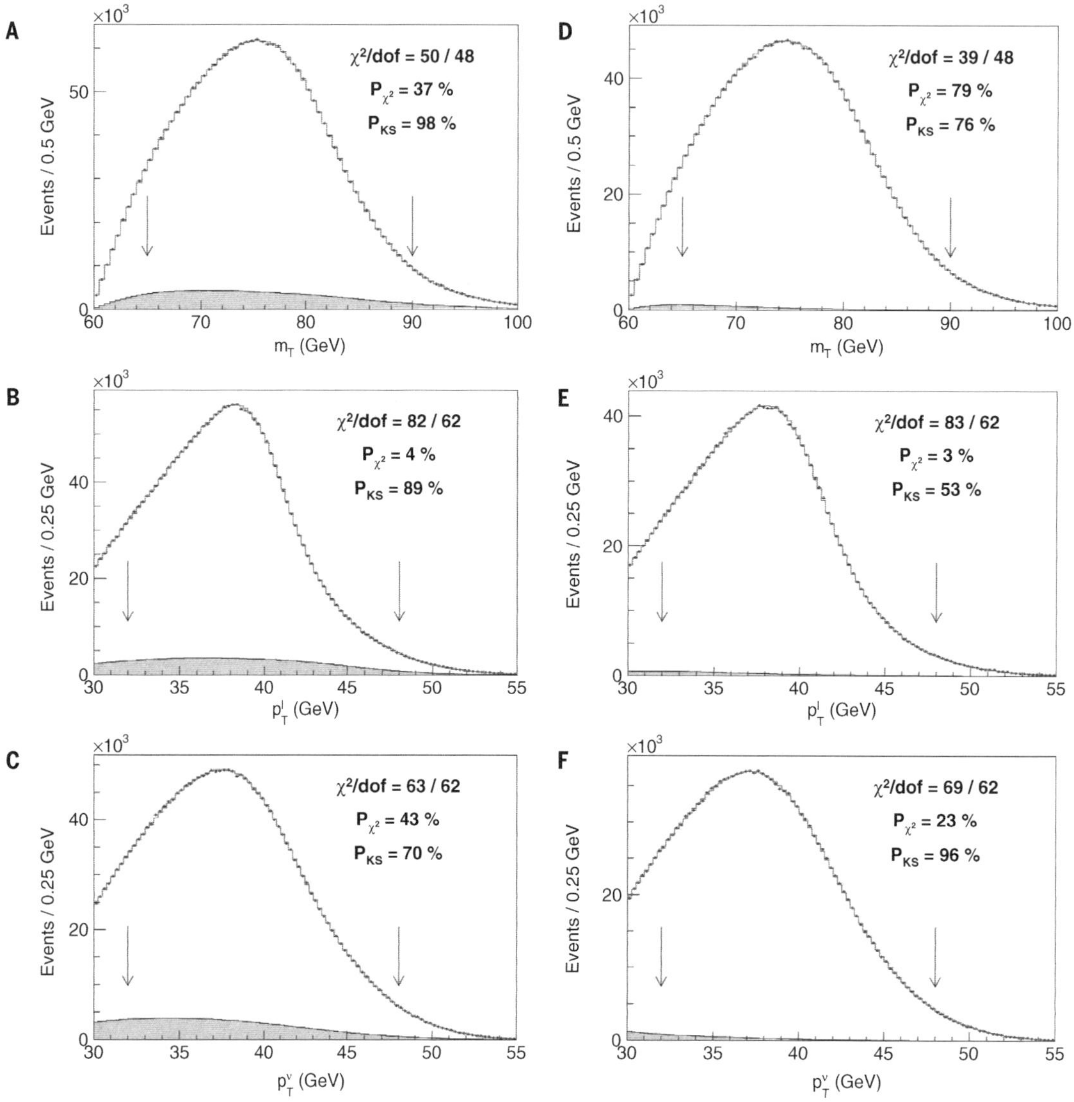

Figure 4. Fits of the simulated signal templates (red histograms) to the data (blue points) for the extraction of the W boson mass. The arrows mark the fitting windows. The legends show the χ^2/dof, the χ^2 probability P_{χ^2}, and the Kolmogorov-Smirnov probability P_{KS} of the fits. The shaded histograms show the backgrounds, which are kept fixed during the fits. (**A**) m_T fit in the muon channel. (**B**) Lepton p_T fit in the muon channel. (**C**) Neutrino p_T fit in the muon channel. (**D**) m_T fit in the electron channel. (**E**) Lepton p_T fit in the electron channel. (**F**) Neutrino p_T fit in the electron channel.

accounts for multi-photon emission from all charged particles in the production and decay process, confirmed the associated systematic uncertainty of 3 MeV on M_W from EM radiation.

The vector boson p_T model was corrected using CDF data, by fitting the non-perturbative parameter g_2 of ResBos (used in the resummation calculation) and the strong coupling parameter α_s to the measured dilepton p_T spectrum, with an uncertainty of 1.8 MeV on M_W. The

Table 2. W boson mass results obtained from the fits of the kinematic distributions.

Distribution	W boson mass (MeV)	χ^2/dof
$m_T(e,\nu)$	$80\,429.1 \pm 10.3_{stat} \pm 8.5_{syst}$	39/48
$p_T^\ell(e)$	$80\,411.4 \pm 10.7_{stat} \pm 11.8_{syst}$	83/62
$p_T^\nu(e)$	$80\,426.3 \pm 14.5_{stat} \pm 11.7_{syst}$	69/62
$m_T(\mu,\nu)$	$80\,446.1 \pm 9.2_{stat} \pm 7.3_{syst}$	50/48
$p_T^\ell(\mu)$	$80\,428.2 \pm 9.6_{stat} \pm 10.3_{syst}$	82/62
$p_T^\nu(\mu)$	$80\,428.9 \pm 13.1_{stat} \pm 10.9_{syst}$	63/62
Combination	$80\,433.5 \pm 6.4_{stat} \pm 6.9_{syst}$	7.4/5

correction was checked by comparing the calculated spectrum of the dilepton opening angle $\phi^*_\eta = \cot(\Delta\phi_{\ell^-\ell^+}/2)\text{sech}(\Delta\eta_{\ell^-\ell^+}/2)$, which in leading order is independent of $p_T^{\ell\ell}$, with the measured spectrum.

Finally, a complete detector model, based on first principles of particle tracking, was developed to simulate all quantities measured in the data. In this model, tracks and photons are propagated through a high resolution 3D lookup table of material properties for the silicon detector and the COT, made from detailed construction-level knowledge. At each material interaction, the ionisation energy loss is calculated according to detailed formulas using the Landau distribution, bremsstrahlung photons are generated down to 0.4 MeV using detailed differential cross section calculations, Compton scattering and photon conversion to e^-e^+ are simulated and then photons and conversion electrons are propagated into the detector, and multiple Coulomb scattering of all charged particles is simulated including non-Gaussian tails. Hits are deposited and smeared on COT wires, and a helix fit for track reconstruction is performed, applying optionally beam constraint. All signals were simulated using this custom MC model. Templates of the fit variables were generated as finely-spaced functions of M_W and then fitted to the data using binned maximum likelihood fits. The fast custom MC model allowed for making smooth, high-statistics templates and provided analysis control over key components of the simulation.

4. Detector studies

The tracking momentum scale was set by using dimuon resonances. The J/ψ mass was fitted in bins of $1/p_T^\mu$ (proportional to the curvature of the muon track) to measure the momentum scale at low p_T^μ. The J/ψ mass was found independent of p_T^μ after 2.6% tuning of the energy loss. Then the Υ mass was fitted inclusively in p_T^μ to measure the momentum scale at higher p_T^μ and to validate the beam-constraining procedure, as Υ decays promptly, by comparing the beam-constrained (BC) and non-beam-constrained (NBC) Υ mass fits. Using the momentum scale extracted from $J/\psi/\Upsilon \to \mu\mu$ data, a blind measurement of M_Z from $Z \to \mu\mu$ data was performed, shown in Fig. 2 **A**, which gave a result of $M_Z = 91\,192.0 \pm 6.4_{stat} \pm 2.3_{mom.\ scale} \pm 3.1_{QED} \pm 1.0_{align.}$ MeV, in excellent agreement with the PDG value of $91\,188 \pm 2$ MeV [3]. Having validated the low p_T^μ calibration, the extrapolation of the momentum scale to high p_T^μ was constrained by fitting the $Z \to \mu\mu$ data using the PDG value for M_Z and then extracting the scale from all $J/\psi/\Upsilon/Z \to \mu\mu$ fits. The final calibration, shown in Fig. 3 **A**, yields a momentum scale correction of $\Delta p/p = -1389 \pm 25$ parts per million, with an uncertainty of 2 MeV on M_W.

For the EM calorimeter response, energy loss distributions were calculated using detailed

Table 3. Uncertainties on the combined W boson mass 2022 result.

Source	Uncertainty (MeV)
Lepton energy scale	3.0
Lepton energy resolution	1.2
Recoil energy scale	1.2
Recoil energy resolution	1.8
Lepton efficiency	0.4
Lepton removal	1.2
Backgrounds	3.3
p_T^Z model	1.8
p_T^W/p_T^Z model	1.3
Parton distributions	3.9
QED radiation	2.7
W boson statistics	6.4
Total	9.4

Table 4. Uncertainties on the combined W boson mass 2012 result.

Source	Uncertainty (MeV)
Lepton energy scale & resolution	7
Recoil energy scale & resolution	6
Lepton removal	2
Backgrounds	3
p_T^W model	5
Parton distributions	10
QED radiation	4
W boson statistics	12
Total	19

GEANT4 calorimeter simulation tuned on CDF data, including leakage into the hadronic calorimeter, absorption in the magnet, and dependence on the incident angle and transverse energy E_T. The energy-dependent gain (non-linearity term) was determined from fits to $W \to e\nu_e$ and $Z \to ee$ E/p data in bins of E_T. The energy scale was determined by fitting the E/p spectrum of the data, as shown in Fig. 3 **B**, which was constructed from the calibrated momentum of the tracks associated to the calorimeter clusters of the selected electrons and from the energies of the clusters. The low tail of the spectrum was used to tune the calorimeter thickness and the high tail to tune the radiative material model in the custom MC simulation. The uncertainty on M_W was 6 MeV from the energy scale and 2 MeV from the non-linearity. A uniformity check was performed in bins of η. Using the energy scale determined from the E/p analysis, a blind measurement of M_Z from $Z \to ee$ data was performed, shown in Fig. 2 **B**, which gave a result of $M_Z = 91\,194.3 \pm 13.8_{stat} \pm 6.5_{calor.} \pm 2.3_{mom.} \pm 3.1_{QED} \pm 0.8_{align.}$ MeV, again in excellent agreement with the PDG value [3]. Having validated the energy calibration with the above procedure, the energy scale was also extracted by fitting the $Z \to ee$ data using the PDG value for M_Z and the two scales were combined for maximum accuracy. The final energy scale correction was $\Delta S_E = -14 \pm 72$ parts per million and the induced uncertainty on M_W was 5.8 MeV.

The tracking resolution was parametrised in the custom simulation by a radius-dependent hit resolution $\sigma_h = 150 \pm 1_{stat}$ μm and a beam spot size $\sigma_b = 36.0 \pm 0.5_{stat}$ μm and was tuned on the widths of the $Z \to \mu\mu$ (BC) and $\Upsilon \to \mu\mu$ (both BC and NBC) mass peaks, giving an intrinsic fractional momentum resolution of $\delta p_T/p_T \simeq 0.0005 p_T/\mathrm{GeV}$ [11]. The uncertainty on M_W from the p_T^μ resolution was 0.3 MeV. The electron cluster resolution was parametrised in the custom simulation by a sampling term of $12.6\%/\sqrt{E_T}$ plus a constant term of $0.73 \pm 0.02_{stat}\%$ and was tuned on the widths of the E/p peak and the $Z \to ee$ mass peak by selecting radiative electrons. The uncertainty on M_W from the E_T resolution was 0.9 MeV.

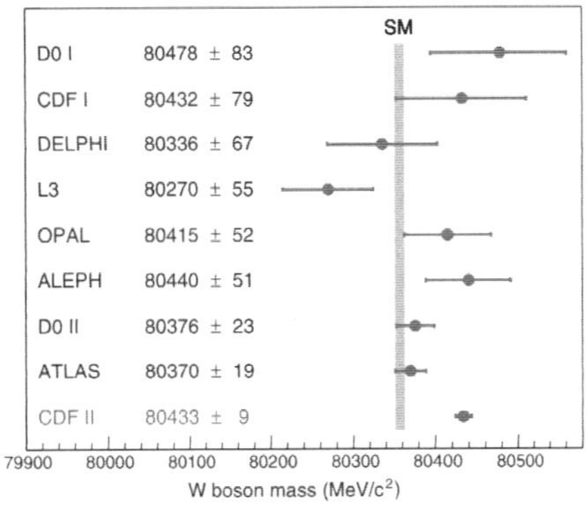

Figure 5. W boson mass measurements compared with the SM expectation (shaded band).

5. Recoil model

The detector response model for the hadronic recoil was derived from data, exploiting the similarity in the production and decay of the W and Z bosons. The model was tuned using the p_T balance in $Z\to\ell\ell$ events. First, calorimeter towers containing lepton energy were removed from the recoil calculation and the lost underlying event (UE) energy after the lepton tower removal was recovered by rotating azimuthal windows in W boson data. Then, the transverse momentum $\vec{u}$ of the hadronic recoil was calculated as a 2D vector sum over calorimeter towers. This vector has a soft, randomly oriented "spectator interaction" component, which was modelled using minimum bias data with tuneable magnitude, and a hard "jet" component directed opposite to the boson p_T. Based on p_T-dependent calorimeter response and resolution parametrisations, the recoil response $R_{rec} = u_{reconstr.}/u_{true}$ was parametrised as a logarithmically increasing function of the boson p_T, motivated by Z boson data. The vector sum of $\vec{p}_T^{\ell\ell}$ and $\vec{u}$ was projected on an axis η parallel to $\vec{p}_T^{\ell\ell}$ and an axis ξ normal to $\vec{p}_T^{\ell\ell}$. The mean and r.m.s. values of the projections, as functions of $p_T^{\ell\ell}$ provide information on the hadronic recoil model parameters, which are tuned by minimising the χ^2 between data and simulation. The recoil components parallel ($u_{||}$) and normal ($u_\perp$) to the lepton momentum were tested with W boson events in the data. The recoil model was additionally constrained with W boson events by using $d\sigma/(dp_T^V dy_V dM_V)$ $(V = W, Z)$ at NNLL + NNLO accuracy from the program DYqT [12] to model the scale variation of the ratio p_T^W/p_T^Z and then using the measured p_T^W spectrum to reduce the scale uncertainty of the model, taking into account the correlation with the hadronic recoil.

6. Backgrounds

The sources of background are listed in Table 1. The main background originates from $Z/\gamma^*\to\mu\mu$ events. The backgrounds from Drell-Yan $Z/\gamma^*\to\ell\ell$ events, where one lepton escapes detection, and from $W\to\tau\nu\to\ell\nu\nu\nu$ events were estimated from the custom simulation. Backgrounds from multi-jet events where one jet mimics a lepton were estimated in the muon channel using a neural-network (NN) discriminant for misidentified and signal muons in the data, and in the electron channel by fitting signal and background templates of track isolation, NN discriminant, and MET to $W\to e\nu$ data. The background from decays-in-flight of low momentum long-lived mesons to muons in the COT, resulting in reconstructed high-p_T tracks, was estimated by fitting track χ^2/dof templates of $Z\to\mu\mu$ (signal) and $W\to\mu\nu$ with large impact parameter (background) events selected from data to $W\to\mu\nu$ candidates, taking into account the contamination from true $W\to\mu\nu$ events with poor impact parameter resolution. Cosmic rays were vetoed with an efficiency $> 99\%$ using a dedicated tracking algorithm, with the remaining background estimated from a previous data sample scaled by the run time to the integrated luminosity ratio.

7. Results

Fig. 4 shows the post-fit signal and background templates compared with the data for all three sensitive variables in both the muon and electron channels. The values of χ^2/dof, the corresponding χ^2 probability, and the Kolmogorov-Smirnov test probability shown in the legends demonstrate the high quality of the fit in all cases. The results from the various fits were combined using the Best Linear Unbiased Estimator (BLUE) algorithm [13], which takes into account the correlations among the measurements which are determined through pseudo-experiments. Table 2 lists the M_W values obtained from all six fits and the final combination. Full consistency among all fits in both channels is observed. Table 3 lists the systematic uncertainties on the combined M_W result. Comparing with the systematic uncertainties of the previous M_W measurement by CDF [14], shown in Table IV, there is substantial improvement in most cases. Uncertainties from data-driven sources scale with the integrated luminosity, as expected, whereas uncertainties from theory, such as vector boson p_T modelling, proton PDF, and EM radiation modelling, are improved by using updated theoretical inputs.

8. Conclusions

The W boson mass measurement is a highly challenging task, therefore slowly progressing, but is now reaching an impressive precision at hadron collider experiments, approaching the level of precision of the SM expectation. This precision allows for a tight test of the internal consistency of the SM. The new CDF measurement of $M_W = 80\,433.5 \pm 6.4_{stat} + 6.9_{syst}$ MeV $= 80\,433.5 \pm 9.4$ MeV is twice as precise as previous measurements, with a total uncertainty of about 1 part in 10 000. This result is in strong tension, at 7.0σ, with the SM expectation of $80\,357 \pm 6$ MeV. It is also in tension, at the level of 2-3σ, with recent M_W results from LHC experiments, as shown in Fig. 5 which summarises the current status of M_W measurements. Higher precision from the LHC is needed to arrive at a firm conclusion, while the tension with the SM suggests the possibility of improvements in the SM calculation or of extension to the SM.

References

[1] Erler J and Schott M 2019 Electroweak Precision Tests of the Standard Model after the Discovery of the Higgs Boson *Preprint* arXiv:1902.05142
[2] Djouadi A 2008 The anatomy of electroweak symmetry breaking: Tome I: The Higgs boson in the Standard Model *Phys. Rept.* **457** 1-216
[3] Zyla P A *et al* (Particle Data Group) The Review of Particle Physics *Prog. Theor. Exp. Phys.* D **2020** 083C01
[4] Lopez-Val D and Robens T A 2014 Δr and the W-boson mass in the singlet extension of the standard model *Phys. Rev.* D **90** 114018
[5] Aaltonen T *et al* (CDF Collaboration) 2022 High-precision measurement of the W boson mass with the CDF II detector *Science* **04** 376:170-176 & supplemental material in the electronic edition
[6] Kotwal A and Hays C 2014 Drift chamber alignment using cosmic rays *Nucl. Instrum. Methods* A **762** 85-99
[7] http://nnpdf.mi.infn.it
[8] Balacs C and Yuan C-P 1997 Soft gluon effects on lepton pairs at hadron colliders *Phys. Rev.* D **56** 5558-5583
[9] Golonka P and Was Z 2006 PHOTOS Monte Carlo: a precision tool for QED corrections in Z and W decays *Eur. Phys. J.* D **45** 97
[10] Carloni Calame C M, Montagna G, Nicrosini O and Vicini A Precision electroweak calculation of the production of high transverse-momentum lepton pairs at hadron colliders *J. High Energy Phys.* **2007** 109
[11] Aaltonen T *et al* (CDF Collaboration) 2012 First Run II Measurement of the W Boson Mass *Phys. Rev.* D **77** 112001 (2008)
[12] Bozzi G, Catani S, Ferrera G, de Florian D and Grazzini M 2009 Transverse momentum resummation: A perturbative study of Z production at the Tevatron *Nucl. Phys.* B **815** 174-197
[13] Lyons L and Gibaut D 1988 How to combine correlated estimates of a single physical quantity *Nucl. Instrum. Methods* A **270** 110-117
[14] Aaltonen T *et al* (CDF Collaboration) 2012 Precise Measurement of the W-Boson Mass with the CDF II Detector *Phys. Rev. Lett.* **108** 151803; *Phys. Rev.* D **89** 072003 (2014)

First search for magnetic monopoles through the Schwinger mechanism

Vasiliki A Mitsou[1,2] **for the MoEDAL Collaboration**

[1] Instituto de Física Corpuscular (IFIC), CSIC – Universitat de València, C/ Catedrático José Beltrán 2, E-46980 Paterna (Valencia), Spain
[2] National and Kapodistrian University of Athens, University Campus, GR-15784 Zografou, Athens, Greece

E-mail: `vasiliki.mitsou@ific.uv.es`

Abstract. Magnetic monopoles are hypothetical fundamental particles predicted in several theories beyond the standard model, however they have never been experimentally detected. The Schwinger mechanism predicts that an extremely strong magnetic field would produce isolated magnetic charges, if they exist. Looking for the existence of magnetic monopoles via the Schwinger mechanism had not been attempted before, but it is advantageous, owing to the possibility of calculating its rate through semi-classical techniques without perturbation theory. This paper focuses on the first search for magnetic monopoles produced by the Schwinger mechanism in heavy-ion collisions. It was carried out by the MoEDAL experiment, whose trapping detectors were exposed to 0.235 nb^{-1} of Pb-Pb collisions with 5.02 TeV energy per collision at the LHC, that provided the strongest known magnetic fields in the universe. A superconducting quantum interference device magnetometer scanned these detectors for the presence of magnetic charge. Magnetic monopoles with 1, 2 and 3 Dirac charges and masses up to 75 GeV were excluded by the analysis. This analysis, which has been published in the journal *Nature*, provided a lower mass limit for finite-size magnetic monopoles from a collider search and greatly extended previous mass bounds.

1. Introduction

Magnetic monopoles are hypothetical fundamental particles predicted in several theories beyond the standard model (SM), however they have never been experimentally detected up to now. In collider experiments, their production through Feyman diagrams such as Drell-Yan-like or photon-fusion mechanisms has been considered in related searches. The exploration of the Schwinger mechanism as a possible means to produce them thermally in strong magnetic fields constitutes a paradigm shift. Such an analysis has been performed recently [1] by the MoEDAL experiment at the Large Hadron Collider (LHC).

The paper is structured as follows. The main features and motivation for magnetic monopole are highlighted in section 2. In section 3, the MoEDAL physics goals are reviewed and its detectors are briefly described. After an overview of the previous MoEDAL searches for monopoles in section 4, the analysis considering the Schwinger thermal production is presented in section 5. Section 6 discusses future prospects for monopole searches among other proposals envisaged by the MoEDAL Collaboration to extend significantly its physics program to monopoles and beyond. The paper concludes with a summary in section 7.

2. Magnetic monopoles

Magnetic monopoles [2] that carry a non-zero magnetic charge and dyons [3] possessing both magnetic and electric charge are among the most fascinating hypothetical particles. Even though there is no generally acknowledged empirical evidence for their existence, there are strong theoretical reasons to believe that they do exist, and they are predicted in many theories.

The theoretical motivation behind the introduction of magnetic monopoles is the symmetrisation of the Maxwell equations and the explanation of the charge quantisation [4]. Dirac showed that the mere existence of a monopole in the universe could offer an explanation of the discrete nature of the electric charge, leading to the Dirac Quantisation Condition (DQC),

$$eg = \frac{N}{2}, \quad N \in \mathbb{Z}, \tag{1}$$

where e is the electron charge and g the monopole magnetic charge. In Dirac's formulation, magnetic monopoles are assumed to exist as point-like particles and quantum mechanical consistency conditions lead to (1), establishing the value of their magnetic charge. Although monopoles symmetrise Maxwell equations in form, there is a numerical asymmetry arising from the DQC, namely that the basic magnetic charge is much larger than the smallest electric charge. A magnetic monopole with a single Dirac charge, g_{D}, has an equivalent electric charge of $e/2\alpha = 137e/2$, where α is the fine-structure constant. Thus for a relativistic monopole the energy loss is around $4,700$ times (68.5^2) larger than that of a minimum-ionising electrically charged particle. The monopole mass and spin remain free parameters of the theory.

The existence of monopoles in grand unified gauge theories (GUTs) has been proposed by 't Hooft [5] and Polyakov [6], yet in this case its mass would be near the GUT scale, therefore outside the reach of any currently operating or any future particle accelerator. Unlike the Dirac monopole, GUT-motivated monopoles are extended objects, the production of which is exponentially suppressed by a factor $e^{-4/\alpha}$. Monopoles also emerge in superstring theory [7, 8] and as self-gravitating *global monopoles* [9–15].

In 1986, Cho and Maison proposed the *electroweak monopole* [16, 17], a non-trivial hybrid between the (Abelian) Dirac and the (non-Abelian) 't Hooft–Polyakov monopole. Its predicted magnetic charge is $2g_{\mathrm{D}}$ and, most importantly, its mass is between 4 to 7 TeV, which is accessible even to the LHC energies [18, 19].

Monopoles can also contribute to light-by-light scattering in colliders through radiative corrections in the form of a *box diagram* [20]. Such loop effects would lead to the production of highly energetic photons, which could provide a handle to detect them indirectly in experiments such as ATLAS and CMS or in future accelerators such as the International Linear Collider and the Future Circular Collider in the context of a Born–Infeld theory [21–24]. This possibility is viable if a point-like monopole is considered; the effect is highly suppressed for finite-monopoles, in which case only the Schwinger mechanism could provide some sensitivity.

The lack of experimental confirmation of free monopoles may be due to the bound state called *monopolium* [25–27] they may form, being confined by strong magnetic forces. Monopolium is a neutral state, hence it is difficult to detect directly at a collider detector, although its decay into photons would give a rather clear signal to the ATLAS and CMS detectors [20, 28–30], unlike MoEDAL that cannot detect photons. Nevertheless, the monopolium might break up into highly ionising dyons, which subsequently can be detected in MoEDAL. Moreover, its decay via photon emission would produce a peculiar trajectory in the medium, if the decaying states are also magnetic multipoles [31]. The monopolium is also a viable dark matter candidate [32].

3. The MoEDAL experiment

The MoEDAL (Monopole and Exotics Detector at the LHC) [33] experiment at the LHC [34] is dedicated to searches for manifestations of new physics through highly ionising particles (HIPs) in a manner complementary to ATLAS and CMS [35]. It is the first dedicated *search* LHC experiment to be approved among others that followed [36–38]. The principal motivation for MoEDAL is the quest for magnetic monopoles and dyons, as well as any massive, metastable, slow-moving particle with the fundamental electric charge, or multiples thereof [39–41], arising in various extensions of the SM [31], such as SUSY long-lived spartners [42–45], D-matter [46–52] among others [31, 53].

The MoEDAL detector [54] is deployed around the intersection region at Point 8 (IP8) of the LHC in the LHCb vertex locator (VELO) cavern. A three-dimensional schematic view of the MoEDAL experiment is presented in figure 1. It is a unique and largely passive detector, which does not require neither readout or trigger, based on three different detection techniques. It provides a permanent physical register of the HIPs passage without the presence of background from SM processes.

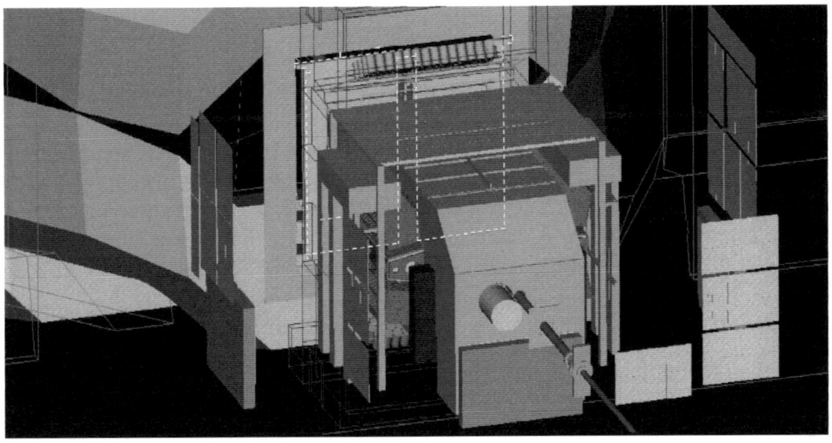

Figure 1. GEANT4 visualisation of the MoEDAL experiment around the LHCb VELO region at Point 8 of the LHC.

The main sub-detector system is made of a large array of CR39®, Makrofol® and Lexan® nuclear track detector (NTD) stacks surrounding the IP8. The passage of a HIP through the plastic sheets is marked by an invisible damage zone along the trajectory. The damage zone is revealed as a cone-shaped etch-pit when the plastic detector is chemically etched at the INFN Bologna laboratory. Then the sheets of plastics are scanned by an automatic system looking for aligned etch pits in multiple sheets.

A unique feature of MoEDAL is the use of magnetic monopole trappers (MMTs) to capture magnetically charged particles. The aluminium absorbers of MMTs are subject to an analysis in a superconducting quantum interference device (SQUID) magnetometer looking for isolated magnetic charges [55]. During the scanning, which takes place at the ETH Zurich SQUID laboratory, the persistent current is measured before and after the passing of the sample through the superconducting loop. A consistent difference between the two currents after several passings of the same sample would provide a clear signal of a trapped magnetic monopole in the aluminium volume.

The only active MoEDAL sub-detector comprises an array of several TimePix pixel devices dedicated to the monitoring of cavern background sources. Its time-over-threshold mode provides a three-dimensional mapping of the charge spreading effect in the silicon sensor volume, thus differentiating between distinct particles species from mixed radiation fields and measuring their energy deposition [56].

4. Previous searches for monopoles and dyons in MoEDAL

MoEDAL is designed to fully exploit the energy-loss mechanisms of magnetically charged particles in order to optimise its potential to discover these elusive particles. The MoEDAL physics results on monopoles are mostly based on the scanning of the MMTs, exposed to LHC Run 1 data at $\sqrt{s} = 8$ TeV [57] and to 13 TeV pp collisions [58–60]. A distribution of persistent current calibrated to magnetic charges is shown in figure 2, obtained in the Run 2 analysis with the full MMT detector [60]. The SQUID analysis yielded no observed isolated magnetic charges, leading to upper limits on monopole production cross sections. This outcome led to lower mass exclusion bounds when considering two pair production processes: (a) a Drell–Yan-like (DY) process in photon s-channel intermediation, and (b) a photon-fusion t-channel diagram [61].

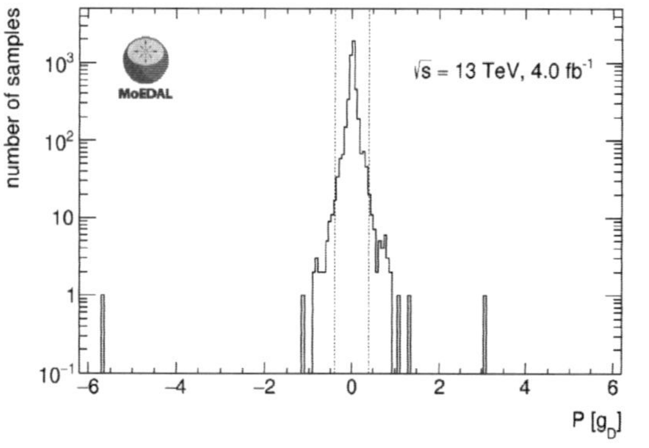

Figure 2. SQUID analysis results: Magnetic pole strength (in units of Dirac charge, $g_{\rm D}$) measured through the induced persistent current in aluminium samples of the MoEDAL trapping detectors exposed to 13 TeV pp collisions with every sample scanned twice. From [60].

MoEDAL pioneered monopole searches in several ways. It considered for the first time spin-1 monopoles [59] and the photon-fusion production mechanism [60]. The dependency of the monopole–photon coupling on $\beta = \sqrt{1 - 4M^2/\hat{s}}$ implies dependency on the centre-of-mass energy of the scattered partons, $\hat{s}$, in terms of the monopole mass, M [61]. This dependency has been consistently taken into account in the MoEDAL results interpretations.

If only the DY production mechanism is considered, the MoEDAL results complement exclusion limits obtained by the ATLAS Collaboration [62–64], which placed limits for monopoles with magnetic charge $|g| \leq 2g_{\rm D}$, as demonstrated in figure 3. The ATLAS bounds, in this case, are better that the MoEDAL ones for low charges due to the higher luminosity delivered to ATLAS than MoEDAL. The acceptance in MoEDAL for small magnetic charges is expected to be enhanced when the —sensitive to low charges— CR39 NTDs be analysed. On the other hand, higher charges are difficult to be probed in ATLAS due to the limitations of the electromagnetic-calorimeter-based level-1 trigger deployed for such searches [64]. MoEDAL is the sole detector sensitive to high magnetic charges.

It should be further noted that ATLAS cannot differentiate monopoles from HECOs since the corresponding analysis is only sensitive to the ionisation signal.[1] In the MoEDAL NTDs, on the other hand, etch-pits from magnetic charges would be different than those from electric charges, providing a clear signal of magnetic monopoles. Limits on monopoles from other collider experiments as well as from searches in cosmic radiation are reviewed in Refs. [2, 67].

A comparison between the photon-fusion and the DY mass limits is presented in figure 4 together with bounds set by ATLAS [62–64]. The production cross section at the LHC energies for photon fusion is much higher than the DY [61]. Therefore MoEDAL, being the only experiment considering it, set the most stringent limits on monopoles overall [69].

[1] A clear monopole-induced signal in ATLAS (or CMS) could come from an analysis based on the characteristic trajectory of a magnetically charged particle in the presence of a solenoid magnet, exploited in the past by the CLEO [65] and TASSO [66] experiments.

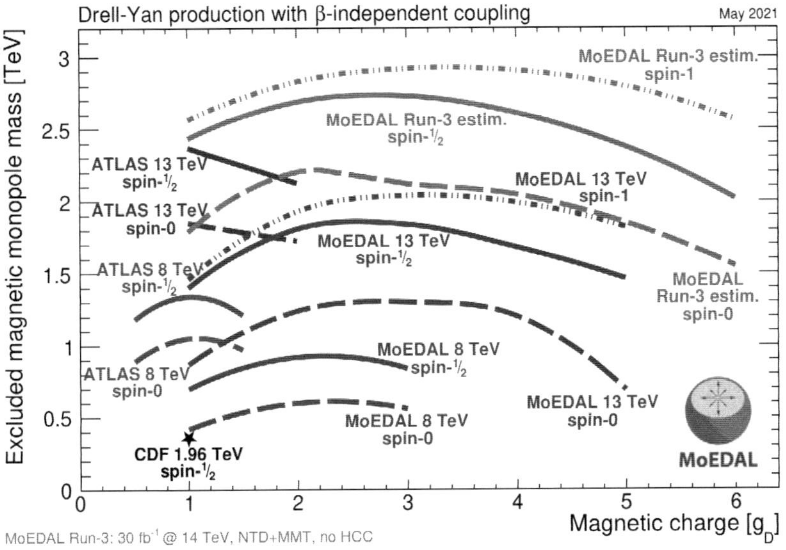

Figure 3. Magnetic monopole mass limits from CDF [68], ATLAS [63, 64] and MoEDAL searches [57, 60] as a function of magnetic charge for various spins, assuming a Drell-Yan pair-production mechanism and β-independent coupling.

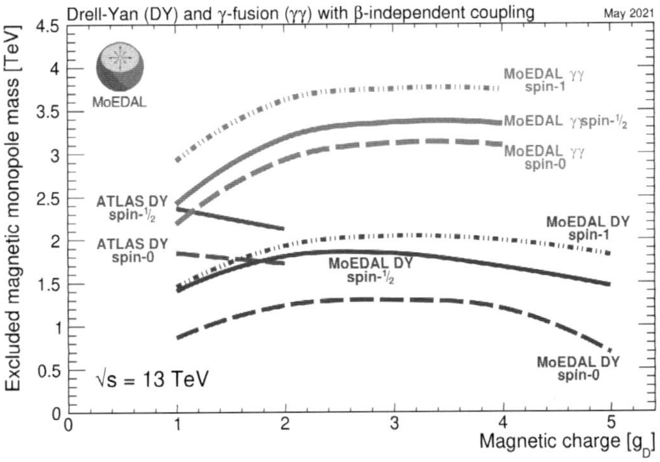

Figure 4. Magnetic monopole mass limits from ATLAS [64] and MoEDAL searches [57, 60] at $\sqrt{s}$ = 13 TeV as a function of magnetic charge for various spins, assuming β-independent coupling and two pair-production mechanism: Drell-Yan and photon fusion. From [69].

MoEDAL performed the first dedicated dyon search [70] in a collider experiment by means of MMT scanning. Mass limits in the range 750–1910 GeV were set using a benchmark DY production model for dyons with magnetic charge up to $5g_{\mathrm{D}}$, for electric charge from $1e$ to $200e$, and for spins 0, 1/2 and 1.

Recently a search for monopoles and for high electrically charged objects (HECOs) involving for the first time NTDs was carried out using a prototype array of Macrofol NTDs exposed to $\sqrt{s} = 8$ TeV pp collisions [71]. The limits placed on HECO pair production, which varied from $\sim$ 30 fb to 70 pb, for electric charges in the range $15e$ to $175e$ and masses from 110 GeV to 1020 GeV, were the most stringent as far as charges is concerned. For comparison, ATLAS has constrained HECOs of electric charge between $20e$ to $100e$ [64] at 13 TeV, an energy higher than that of the MoEDAL analysis. The MoEDAL result was achieved by analysing the Macrofol plastics, characterised by a high detection threshold of $z/\beta = 50$; the, not analysed yet, CR39 plastics are sensitive to charges as low as $z/\beta = 5$. Hence, MoEDAL has a much larger discovery potential, as discussed in section 6.

In both production processes, the monopole pair couples to the photon via a coupling that depends on g_D and therefore has a value of $\mathcal{O}(10)$. This large monopole–photon coupling invalidates any perturbative treatment of the cross-section calculation and hence any result based on it is *only indicative* and used merely to facilitate comparisons between experiments. It is emphasised, however, that the upper bounds placed on production cross sections are solid and can be relied upon. One way to resolve the problem of cross-section calculations with large couplings is to use resummation techniques [72].

5. Search for monopoles produced by the Schwinger mechanism

Electrically charged particles can be created in strong enough electric fields, a phenomenon known as the Schwinger mechanism [73]. By electromagnetic duality, a sufficiently strong magnetic field would similarly produce magnetic monopoles. The possibility of calculating its rate through semi-classical techniques [74–79] overcomes the non-perturbativity of the monopole–photon coupling.

Heavy-ion collisions at the LHC produce the strongest known magnetic fields in the universe, and the first search for such production was conducted by MoEDAL during the 5.02 TeV/nucleon heavy-ion run, during which the MMTs were exposed to 0.235 nb^{-1} of Pb-Pb collisions. The strongest fields are generated in ultraperipheral collisions, for which the impact parameter b is approximately twice the nuclear radius; a schematic view of the collisions is shown in figure 5. In the 2018 heavy-ion run at the LHC, the peak magnetic field strength was $B \approx 10^{16}$ T [80]. This field strength is about seven orders of magnitude greater than the critical field strength of quantum electrodynamics, and more than four orders of magnitude greater than the strongest known astrophysical magnetic fields, which are present on the surfaces of magnetars [81].

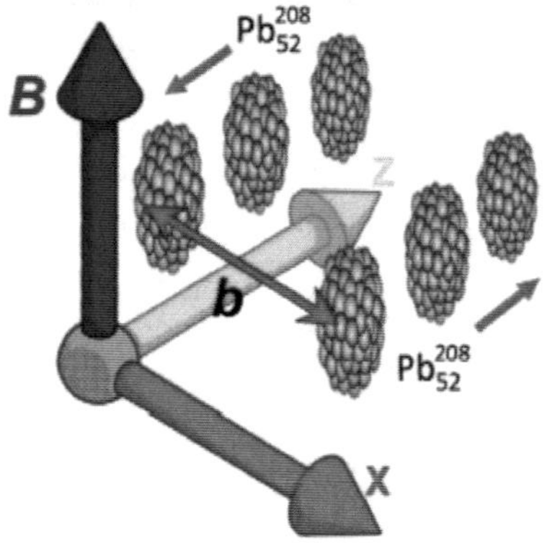

Figure 5. Peripheral Pb-Pb heavy-ion collisions produce strong magnetic fields. The 2018 Pb-Pb run with an integrated luminosity at IP8 of 0.235 nb^{-1}. B is the magnetic field of the heavy ions along the y axis, the z axis is the direction of motion of ions, and the x axis is the direction along which the impact parameter b is measured. From [1].

After exposure during November 2018, the MMTs were analysed with the SQUID at ETH Zurich. The measurements were compatible with the absence of monopoles and therefore magnetic charges equal to or above the Dirac charge were excluded in all samples [1], in a manner similar to that shown in figure 2 for a previous MoEDAL MMT analysis [60]. Subsequently, upper limits on the production cross section at 95% confidence level (CL) were set for monopoles with Dirac charges 1, 2 and $3g_D$, as shown in figure 6.

Two approximations in the calculation of the monopole production cross section have been considered [1].

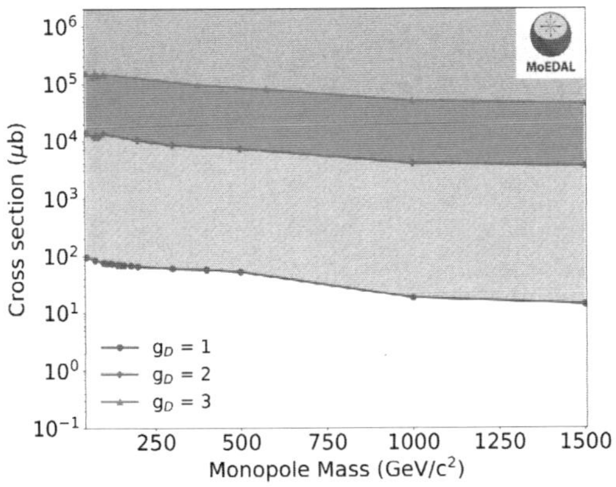

Figure 6. 95% CL exclusion regions on the cross section for monopole production via the Schwinger mechanism as functions of the monopole mass for magnetic charges $1g_{\mathrm{D}}$ (blue), $2g_{\mathrm{D}}$ (red), and $3g_{\mathrm{D}}$ (green). From [1].

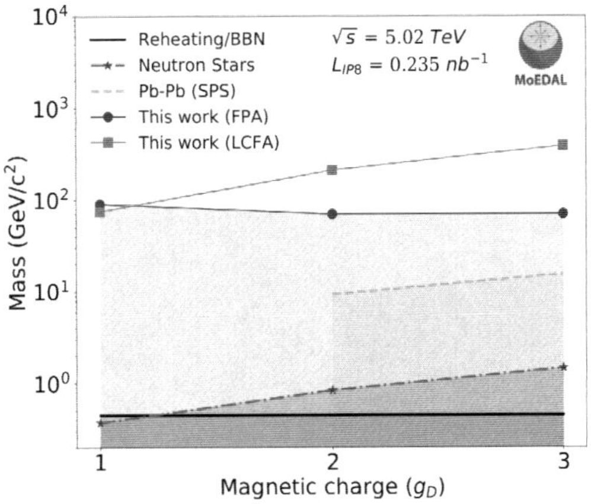

Figure 7. 95% CL excluded monopole masses obtained using the FPA (blue) and LCFA (red) calculations of Schwinger production with the conservative exclusion region shaded violet. Previous limits [74] are also shown for comparison. From [1].

Free-particle approximation (FPA) The spacetime dependence of the electromagnetic field of the heavy ions is treated exactly, but monopole self-interactions are neglected [79].

Locally constant field approximation (LCFA) The spacetime dependence of the electromagnetic field is neglected, but monopole self-interactions are treated exactly [77].

In this way the two approaches are complementary, with uncorrelated uncertainties. In addition, for the FPA the leading effects of monopole self-interactions have been shown to enhance the cross section, and for the LCFA the leading effects of spacetime dependence have also been shown to increase the cross section [77,79]. Thus, while neither approximation provides a complete calculation of the production cross section, both are expected to yield conservative lower limits. Thus, monopoles with Dirac charges $1g_{\mathrm{D}} \leq g \leq 3g_{\mathrm{D}}$ and masses up to 75 GeV were excluded, as seen in figure 7. The mass reach appears to be 20–30 times lower than the current bounds from ATLAS and MoEDAL (cf. figure 4), however, this cross-section calculation is theoretically sound. This analysis provided the first lower mass limits for finite-size monopoles from a collider search [1].

6. Future prospects

The MoEDAL Collaboration plans to continue scanning MMTs exposed to heavy-ion collisions both in the LHC Run 3 [82] and in the subsequent high-luminosity Run 4 (HL-LHC) looking for thermally produced monopoles. Assuming 2.5 nb^{-1} Pb-Pb collisions at $\sqrt{s_{\mathrm{NN}}} = 5.52$ TeV and conservative theoretical assumptions, a ~ 20 GeV increase in sensitivity in HL-LHC heavy-ion run is expected, as observed in figure 8. Nuclear track detectors are not included in this projection [83].

MMTs exposed to pp collisions will also be analysed for the presence of monopoles during Run 3 and Run 4 assuming DY and photon-fusion production mechanisms. Likewise, NTDs will be etched and scanned looking for traces of magnetically and/or electrically charged particles. Actually, such an analysis for NTDs exposed to Run 2 collisions is currently underway, following the one presenting NTD and MMT results for Run 1 [71]. Processing CR39 plastic sheets, besides

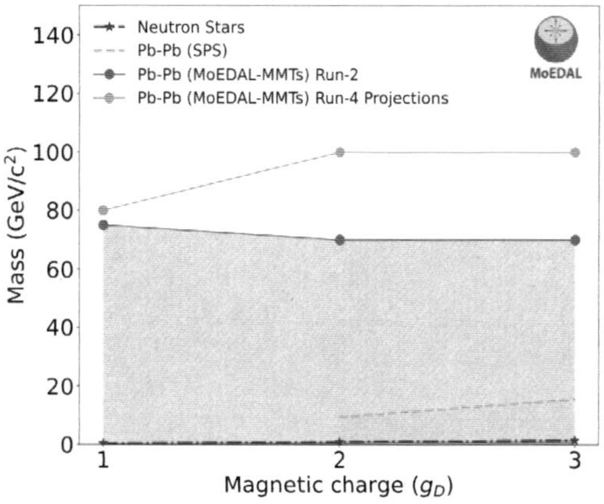

Figure 8. Projected 95% CL mass exclusion region (magenta line) expected for HL-LHC Pb-Pb runs using a conservative Schwinger cross-section calculation compared to the MoEDAL search limits (green line) [1]. From [83].

Macrofol, will allow to probe relatively *low* electric charges of less than $10e$ [41]. Moreover, an analysis for trapped monopoles in the Run 1 CMS beam pipe is ongoing [84].

MoEDAL plans to extend its physics program to *feebly interacting particles* that connect hidden sectors to the visible SM sector with the MoEDAL Apparatus for Penetrating Particles (MAPP) [85, 86]. Such *portal* scenarios attempt to explain observed phenomena in particle physics, astrophysics and cosmology such as the non-zero neutrino masses and oscillations, the dark matter [87, 88] and the baryon asymmetry of the universe, among others. MAPP Phase 1 was approved by the CERN Research Board in 2021 [89] and it is currently being deployed in the UA83 tunnel some 100 m from IP8, protected by cosmic rays by an overburden of $\sim$ 100 m of limestone [90]. MAPP Phase 2 is going to be in operation in HL-LHC and extend MAPP-1 physics reach [**?**].

Besides the search for monopoles in experiments with colliding beams, the MoEDAL Collaboration is also considering [85, 91] a *monopole telescope* sensitive to masses up to the GUT scale. This "Cosmic-MoEDAL", which will comprise a 10,000 m^2 array of plastic NTDs deployed at high altitude, will have flux sensitivity well below the Parker limit.

7. Summary

MoEDAL, the first dedicated search experiment at the LHC, is extending considerably the reach in the search for (meta)stable highly ionising particles. It has pioneered several aspects of magnetic monopoles, such as the consideration of vector monopoles, β-dependent coupling and photon-fusion production. Thanks to its unique passive detectors, it provides sensitivity complementary to that of the LHC main detectors. It is the only contender in high magnetic charges and performed the first dedicated collider search for dyons. Very high electric charges have been probed recently by analysing plastic nuclear track detectors.

A breakthrough analysis bypassing the issue of the non-perturbativity of the monopole-photon coupling by considering the Schwinger thermal mechanism for the monopole production in strong magnetic fields has been published recently in *Nature*. It involves the monopole trappers exposure to heavy-ion LHC collisions and yielded the first theoretically robust lower mass limits for monopoles with magnetic charge up to three Dirac charges.

The search for trapped monopoles in the Run 1 CMS beam pipe is expected to be released soon. MoEDAL is also sensitive to single and multiple electric charges predicted in supersymmetric models and other exotic scenarios. Moreover, the search for HIPs can be pushed to GUT mass scales by the deployment of the Cosmic-MoEDAL array at high altitude.

Acknowledgments
The author acknowledges support by the Generalitat Valenciana via a special grant for MoEDAL, via the Excellence Grant Prometeo CIPROM/2021/073 and via the mobility grant CIBEST/2021/29, by the Spanish MICINN/AEI and the European-Union/FEDER via the grants PGC2018-094856-B-I00 and PID2021-122134NB-C21, and by the CSIC AEPP2021 grant 2021AEP063.

References
[1] Acharya B *et al.* (MoEDAL) 2022 *Nature* **602** 63–67 (*Preprint* 2106.11933)
[2] Mavromatos N E and Mitsou V A 2020 *Int. J. Mod. Phys. A* **35** 2030012 (*Preprint* 2005.05100)
[3] Schwinger J S 1969 *Science* **165** 757–761
[4] Dirac P A M 1931 *Proc. Roy. Soc. Lond. A* **133** 60–72
[5] 't Hooft G 1974 *Nucl. Phys. B* **79** 276–284
[6] Polyakov A M 1974 *JETP Lett.* **20** 194–195 [Pisma Zh.Eksp.Teor.Fiz. 20 (1974) 430-433]
[7] Harvey J A and Liu J 1991 *Phys. Lett. B* **268** 40–46
[8] Lazarides G, Panagiotakopoulos C and Shafi Q 1987 *Phys. Rev. Lett.* **58** 1707
[9] Barriola M and Vilenkin A 1989 *Phys. Rev. Lett.* **63** 341
[10] Drukier A K and Nussinov S 1982 *Phys. Rev. Lett.* **49** 102
[11] Mazur P O and Papavassiliou J 1991 *Phys. Rev. D* **44** 1317–1320
[12] Mavromatos N E and Papavassiliou J 2018 *Eur. Phys. J. C* **78** 68 (*Preprint* 1712.03395)
[13] Mavromatos N E and Sarkar S 2017 *Phys. Rev. D* **95** 104025 (*Preprint* 1607.01315)
[14] Mavromatos N E and Sarkar S 2018 *Phys. Rev. D* **97** 125010 (*Preprint* 1804.01702)
[15] Mavromatos N E and Sarkar S 2018 *Universe* **5** 8 (*Preprint* 1812.00495)
[16] Cho Y M and Maison D 1997 *Phys. Lett. B* **391** 360–365 (*Preprint* hep-th/9601028)
[17] Bae W S and Cho Y M 2005 *J. Korean Phys. Soc.* **46** 791–804 (*Preprint* hep-th/0210299)
[18] Ellis J, Mavromatos N E and You T 2016 *Phys. Lett. B* **756** 29–35 (*Preprint* 1602.01745)
[19] Ellis J, Hung P Q and Mavromatos N E 2021 *Nucl. Phys. B* **969** 115468 (*Preprint* 2008.00464)
[20] Epele L N, Fanchiotti H, Canal C A G, Mitsou V A and Vento V 2012 *Eur. Phys. J. Plus* **127** 60 (*Preprint* 1205.6120)
[21] Ellis J, Mavromatos N E and You T 2017 *Phys. Rev. Lett.* **118** 261802 (*Preprint* 1703.08450)
[22] Ellis J, Mavromatos N E, Roloff P and You T 2022 *Eur. Phys. J. C* **82** 634 (*Preprint* 2203.17111)
[23] Musumeci E and Mitsou V A 2022 *PoS* **ICHEP2022** to appear
[24] Mitsou V A and Musumeci E 2022 Constraining magnetic monopoles in photon final states at colliders, in preparation
[25] Hill C T 1983 *Nucl. Phys. B* **224** 469–490
[26] Dubrovich V K 2002 *Grav. Cosmol. Suppl.* **8N1** 122–125
[27] Epele L N, Fanchiotti H, Garcia Canal C A and Vento V 2008 *Eur. Phys. J. C* **56** 87–95 (*Preprint* hep-ph/0701133)
[28] Epele L N, Fanchiotti H, Canal C A G and Vento V 2009 *Eur. Phys. J. C* **62** 587–592 (*Preprint* 0809.0272)
[29] Vento V 2018 *Universe* **4** 117
[30] Vento V and Traini M 2020 *Eur. Phys. J. C* **80** 62 (*Preprint* 1909.03952)
[31] Acharya B *et al.* (MoEDAL) 2014 *Int. J. Mod. Phys. A* **29** 1430050 (*Preprint* 1405.7662)
[32] Vento V 2021 *Eur. Phys. J. C* **81** 229 (*Preprint* 2011.10327)
[33] Pinfold J *et al.* (MoEDAL) 2009 Technical Design Report of the MoEDAL Experiment CERN-LHCC-2009-006, MoEDAL-TDR-001
[34] Evans L and Bryant P 2008 *JINST* **3** S08001
[35] De Roeck A, Katre A, Mermod P, Milstead D and Sloan T 2012 *Eur. Phys. J. C* **72** 1985 (*Preprint* 1112.2999)
[36] Alimena J *et al.* 2020 *J. Phys. G* **47** 090501 (*Preprint* 1903.04497)
[37] Mitsou V A 2021 *PoS* **LHCP2020** 112
[38] Mitsou V A 2021 *16th Marcel Grossmann Meeting on Recent Developments in Theoretical and Experimental General Relativity, Astrophysics and Relativistic Field Theories* (*Preprint* 2111.03036)
[39] Hirsch M, Masełek R and Sakurai K 2021 *Eur. Phys. J. C* **81** 697 (*Preprint* 2103.05644)
[40] Masełek R, Altakach M M, Lamba P, Sakurai K and Mitsou V A 2022 *PoS* **DISCRETE2020-2021** 081
[41] Altakach M M, Lamba P, Masełek R, Mitsou V A and Sakurai K 2022 *Eur. Phys. J. C* **82** 848 (*Preprint* 2204.03667)
[42] Fairbairn M, Kraan A C, Milstead D A, Sjostrand T, Skands P Z and Sloan T 2007 *Phys. Rept.* **438** 1–63 (*Preprint* hep-ph/0611040)

[43] Sakurai K, Felea D, Mamuzic J, Mavromatos N E, Mitsou V A, Pinfold J L, Ruiz de Austri R, Santra A and Vives O 2020 *J. Phys. Conf. Ser.* **1586** 012018 (*Preprint* 1903.11022)
[44] Felea D, Mamuzic J, Masełek R, Mavromatos N E, Mitsou V A, Pinfold J L, Ruiz de Austri R, Sakurai K, Santra A and Vives O 2020 *Eur. Phys. J. C* **80** 431 (*Preprint* 2001.05980)
[45] Acharya B S, De Roeck A, Ellis J, Ghosh D K, Masełek R, Panizzo G, Pinfold J L, Sakurai K, Shaa A and Wall A 2020 *Eur. Phys. J. C* **80** 572 (*Preprint* 2004.11305)
[46] Shiu G and Wang L T 2004 *Phys. Rev. D* **69** 126007 (*Preprint* hep-ph/0311228)
[47] Ellis J R, Mavromatos N E and Nanopoulos D V 2000 *Gen. Rel. Grav.* **32** 943–958 (*Preprint* gr-qc/9810086)
[48] Ellis J R, Mavromatos N E and Nanopoulos D V 2008 *Phys. Lett. B* **665** 412–417 (*Preprint* 0804.3566)
[49] Ellis J R, Mavromatos N E and Westmuckett M 2004 *Phys. Rev. D* **70** 044036 (*Preprint* gr-qc/0405066)
[50] Ellis J R, Mavromatos N E and Westmuckett M 2005 *Phys. Rev. D* **71** 106006 (*Preprint* gr-qc/0501060)
[51] Mavromatos N E, Sarkar S and Vergou A 2011 *Phys. Lett. B* **696** 300–304 (*Preprint* 1009.2880)
[52] Mavromatos N E, Mitsou V A, Sarkar S and Vergou A 2012 *Eur. Phys. J. C* **72** 1956 (*Preprint* 1012.4094)
[53] Mavromatos N E and Mitsou V A (MoEDAL) 2017 *EPJ Web Conf.* **164** 04001 (*Preprint* 1612.07012)
[54] Pinfold J L 2019 *Universe* **5** 47
[55] De Roeck A, Hächler H P, Hirt A M, Joergensen M D, Katre A, Mermod P, Milstead D and Sloan T 2012 *Eur. Phys. J. C* **72** 2212
[56] Bergmann B *et al.* (MoEDAL) 2021 *PoS* **ICHEP2020** 720
[57] Acharya B *et al.* (MoEDAL) 2016 *JHEP* **08** 067 (*Preprint* 1604.06645)
[58] Acharya B *et al.* (MoEDAL) 2017 *Phys. Rev. Lett.* **118** 061801 (*Preprint* 1611.06817)
[59] Acharya B *et al.* (MoEDAL) 2018 *Phys. Lett. B* **782** 510–516 (*Preprint* 1712.09849)
[60] Acharya B *et al.* (MoEDAL) 2019 *Phys. Rev. Lett.* **123** 021802 (*Preprint* 1903.08491)
[61] Baines S, Mavromatos N E, Mitsou V A, Pinfold J L and Santra A 2018 *Eur. Phys. J. C* **78** 966 [Erratum: Eur.Phys.J.C 79, 166 (2019)] (*Preprint* 1808.08942)
[62] Aad G *et al.* (ATLAS) 2012 *Phys. Rev. Lett.* **109** 261803 (*Preprint* 1207.6411)
[63] Aad G *et al.* (ATLAS) 2016 *Phys. Rev. D* **93** 052009 (*Preprint* 1509.08059)
[64] Aad G *et al.* (ATLAS) 2020 *Phys. Rev. Lett.* **124** 031802 (*Preprint* 1905.10130)
[65] Gentile T *et al.* (CLEO) 1987 *Phys. Rev. D* **35** 1081
[66] Braunschweig W *et al.* (TASSO) 1988 *Z. Phys. C* **38** 543
[67] Patrizii L, Sahnoun Z and Togo V 2019 *Phil. Trans. Roy. Soc. Lond. A* **377** 20180328
[68] Abulencia A *et al.* (CDF) 2006 *Phys. Rev. Lett.* **96** 201801 (*Preprint* hep-ex/0509015)
[69] Mitsou V A (MoEDAL) 2022 *PoS* **EPS-HEP2021** 704 (*Preprint* 2111.03468)
[70] Acharya B *et al.* (MoEDAL) 2021 *Phys. Rev. Lett.* **126** 071801 (*Preprint* 2002.00861)
[71] Acharya B *et al.* (MoEDAL) 2022 *Eur. Phys. J. C* **82** 694 (*Preprint* 2112.05806)
[72] Alexandre J and Mavromatos N E 2019 *Phys. Rev. D* **100** 096005 (*Preprint* 1906.08738)
[73] Schwinger J S 1951 *Phys. Rev.* **82** 664–679
[74] Gould O and Rajantie A 2017 *Phys. Rev. Lett.* **119** 241601 (*Preprint* 1705.07052)
[75] Gould O and Rajantie A 2017 *Phys. Rev. D* **96** 076002 (*Preprint* 1704.04801)
[76] Gould O, Mangles S, Rajantie A, Rose S and Xie C 2019 *Phys. Rev. A* **99** 052120 (*Preprint* 1812.04089)
[77] Gould O, Ho D L J and Rajantie A 2019 *Phys. Rev. D* **100** 015041 (*Preprint* 1902.04388)
[78] Ho D L J and Rajantie A 2020 *Phys. Rev. D* **101** 055003 (*Preprint* 1911.06088)
[79] Gould O, Ho D L J and Rajantie A 2021 *Phys. Rev. D* **104** 015033 (*Preprint* 2103.14454)
[80] Huang X G 2016 *Rept. Prog. Phys.* **79** 076302 (*Preprint* 1509.04073)
[81] Kaspi V M and Beloborodov A 2017 *Ann. Rev. Astron. Astrophys.* **55** 261–301 (*Preprint* 1703.00068)
[82] Pinfold J *et al.* (MoEDAL) 2021 MoEDAL Run-3 Technical Proposal CERN-LHCC-2021-006, LHCC-P-017
[83] d'Enterria D *et al.* 2022 *2022 Snowmass Summer Study* (*Preprint* 2203.05939)
[84] De Roeck A, Mermod P, Pinfold J *et al.* 2017 Searching for trapped magnetic monopoles in LHC accelerator material https://indico.cern.ch/event/623746/attachments/1427507/2190853/expression_of_interest_beam_pipe.pdf
[85] Pinfold J L 2019 *Phil. Trans. Roy. Soc. Lond. A* **377** 20190382
[86] Staelens M A 2021 *Physics From Beyond the Standard Model: Exotic Matter Searches at the LHC with the MoEDAL-MAPP Experiment* Ph.D. thesis Alberta U.
[87] Mitsou V A 2013 *Int. J. Mod. Phys. A* **28** 1330052 (*Preprint* 1310.1072)
[88] Mitsou V A 2019 *15th Marcel Grossmann Meeting on Recent Developments in Theoretical and Experimental General Relativity, Astrophysics, and Relativistic Field Theories* (*Preprint* 1903.11589)
[89] Pinfold J *et al.* (MoEDAL) 2021 MAPP Phase-1 Technical Proposal CERN-LHCC-2021-024, LHCC-P-022
[90] Mitsou V A (MoEDAL) 2022 *PoS* **DISCRETE2020-2021** 017
[91] Pinfold J (ATLAS) 2017 *EPJ Web Conf.* **145** 10001

Journal of Physics: Conference Series **2375** (2022) 012003 doi:10.1088/1742-6596/2375/1/012003

The CMS Trigger System

Milos Dordevic on behalf of the CMS Collaboration

Vinca Institute of Nuclear Sciences, National Institute of the Republic of Serbia, University of Belgrade, Mike Petrovica Alasa 12-14, 11351 Vinca, Belgrade, Serbia

E-mail: milos.djordjevic@cern.ch

Abstract.
The CMS experiment at CERN uses a two-stage triggering system composed of the Level-1 (L1), instrumented with custom-designed hardware boards with an output rate of 100 kHz, and the High Level Trigger (HLT), streamlined version of the offline software reconstruction that runs on the computing farm, allowing to store around 1.5 kHz of rate. New trigger algorithms and new features, as well as optimized trigger menus at both L1 and HLT are mandatory in order to be able to successfully record the events at higher data loads due to increasing luminosity and pileup at the LHC in Run 3. Many measurements and searches will profit from the updates implemented in the CMS trigger. The highlights of Run 2 CMS trigger results will be presented, together with the improvements for Run 3.

1. Introduction

The Compact Muon Solenoid (CMS) is a general purpose experiment [1] built at the Large Hadron Collider (LHC) at CERN. It is made of several sub-detector layers, comprising a silicon tracker, a homogeneous electromagnetic calorimeter (ECAL) and a sampling hadron calorimeter (HCAL) being immersed in a magnetic field of 3.8 T produced by a superconducting solenoid. Outside of the solenoid volume there are muon sub-detectors positioned in a flux-return yoke. CMS uses Particle Flow algorithm [2] for event reconstruction with all sub-detector information.

The circulating proton bunches at the LHC are time spaced with 25 ns intervals in two beams running in opposite directions. These proton bunches collide at 40 MHz rate, but only about 1 kHz was kept for the data analysis at the end of Run 2. The limitation factors to record more data online are the available storage capacities and also computing resources for the data processing. The CMS Trigger System provides a selective readout of data in real time, keeping only events of interest for further physics analysis, aiming at an efficient decision within the rate constraints.

2. The CMS Trigger System design, architecture and implementation

The CMS Trigger System is organised in two tiers, the Level-1 trigger (L1) based on custom-made electronics that is reducing the event rate to about 100 kHz with 3.8 μs latency, and the High Level Trigger (HLT) that is filtering the events with a software running on computing farm based on the commercial CPUs, which further reduced the rate to about 1 kHz in the Run 2. Heterogeneous computing resources are allocated for Run 3 using CPU and GPU reconstruction.

Figure 1 shows a diagram of the upgraded CMS Level-1 trigger system during Run 2. At L1, each event is analysed by the Muon and Calorimeter trigger. The Muon trigger consists of three muon detection systems used early in the processing chain of the trigger, in order to improve the efficiency and resolution, but also the trigger rate. The Calorimeter trigger is used for reconstructing electrons, photons, tau candidates, jets and energy sums. No tracking readout is used at L1 trigger stage. The Global Trigger combines the various objects provided by the Global Muon Trigger and Layer 2 Calorimeter Trigger. A large set of around 400 requirements is performed on the trigger objects, applied in a logical "OR", comprising the L1 trigger menu. At L1 trigger, the reconstruction of electrons and photons is performed using the cluster shape and electromagnetic fraction information to discriminate against jets. The reconstruction of jets at L1 trigger is done using a sliding window algorithm that looks for trigger tower seeds with an energy over a given threshold. The 9×9 trigger towers are summed in order to match the offline reconstructed jets of the cone size $\Delta R = 0.4$ after which the jets are pileup-subtracted and calibrated. The H_T variable is calculated by summing the jet energies with restriction to pseudorapidity η, while the missing E_T variable at L1 trigger is calculated by summing all trigger towers over the η and pileup dependent energy threshold E_T in the full η range. The muons are reconstructed using an extrapolation based track finding in the barrel region, and a pattern based track finding in the overlap and endcap region, where also the BDT regression algorithms are applied to improve the reconstruction [3].

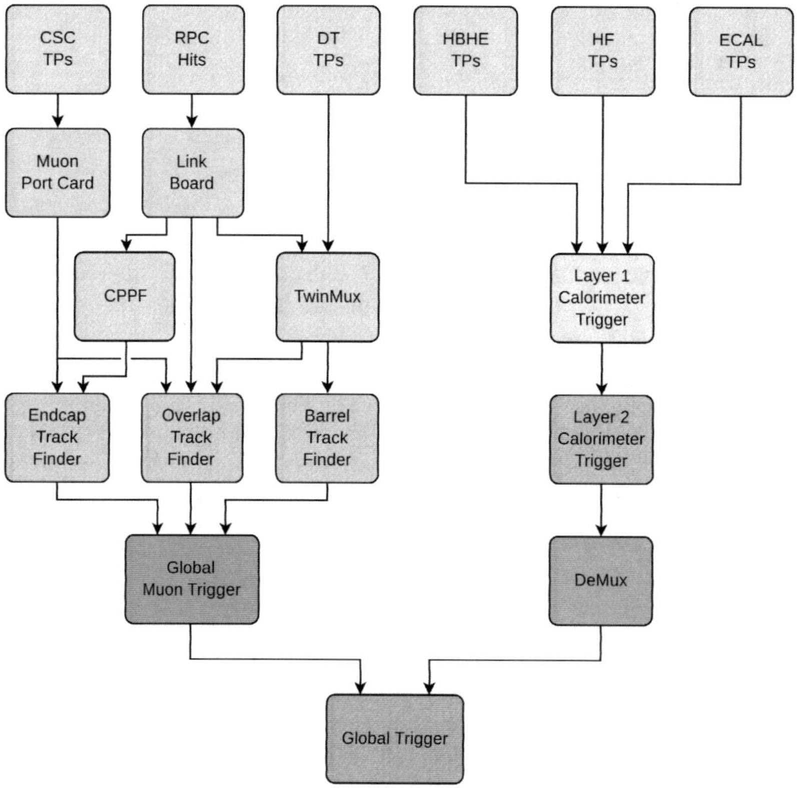

Figure 1. Diagram of the upgraded CMS Level-1 trigger system during Run 2 [3].

3. Performance of the Level-1 Trigger System

The selection of results demonstrating the performance of the L1 trigger in reconstructing of the most relevant physics objects is outlined here. Figure 2 (left) presents the efficiency to reconstruct muons at L1 trigger, as a function of the offline reconstructed muon transverse momentum, shown for three different η ranges and also inclusive up to 2.4 [3]. The efficiency is sharpest in the barrel due to a better resolution for reconstruction of muons in the barrel region. Figure 2 (right) shows the efficiency as a function of the pseudorapidity of the offline reconstructed muon, that exhibits a moderate drop in the forward region. This is due to the fact that in the forward region, corresponding to the EMTF part of the L1 trigger, no detector redundancy is available, as only the Cathode Strip Chambers (CSC) out of the three muons systems of CMS are used. The assignment of muon transverse momentum (p_T), is more difficult due to reduced lever arm and more showering that occurs in the forward region of the detector. The L1 efficiency of the muon track finder is shown to have a small dependence on the number of offline vertices, presenting it to be stable against pileup, and also to have a flat distribution versus the azimuthal angle ϕ [3].

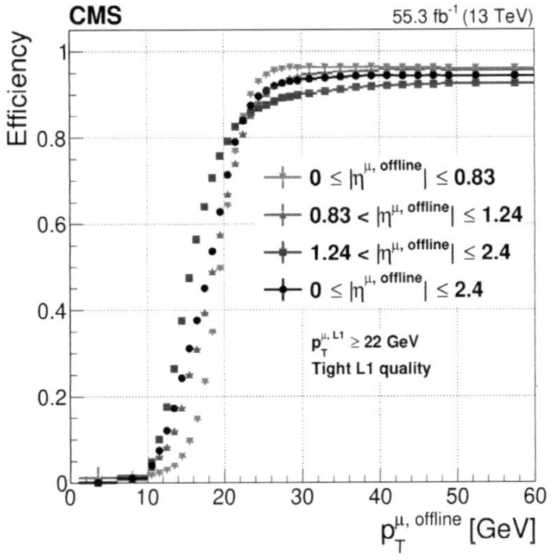

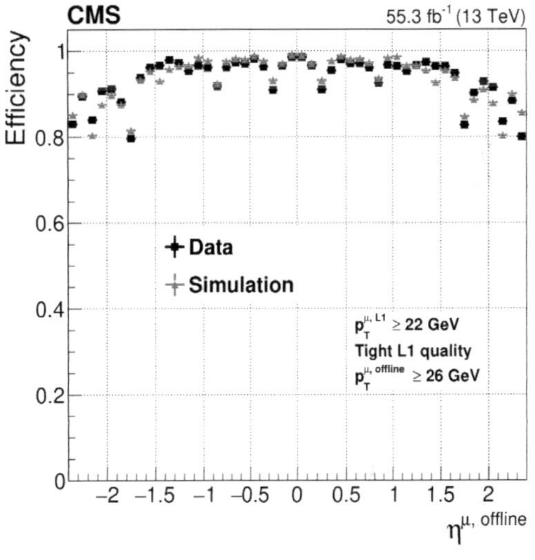

Figure 2. The left plot shows the L1 muon trigger efficiency for data as a function of the offline reconstructed muon $p_T^{\mu,offline}$ for each η region: barrel (red), overlap (violet), endcap (blue) and the total (black) [16]. The right plot shows the L1 muon efficiency for data and simulation as a function of the offline reconstructed muon pseudorapidity [16].

Figure 3 presents the sharp L1 trigger efficiency for reconstructing electrons or photons over given online threshold of 30 and 40 GeV, with respect to the offline reconstructed electron transverse energy [3]. Figure 4 shows only a small dependence of the electron and photon reconstruction and isolation requirements at the L1 with respect to the number of vertices (pileup) [3].

In Figure 5 the single tau object trigger efficiency is presented for the different working points of 26, 30 and 34 GeV for the inclusive tau trigger at the L1 [3]. The energy of the tau leptons is calibrated at the L1 as a function of energy and pileup. The resolution of the tau lepton transverse momentum reconstructed at the L1 trigger is at about 20 to 25 % in the range of 30 to 50 GeV. Figure 6 presents the efficiency curves for the L1 jet triggers for several characteristic thresholds of 35, 90, 120 and 180 GeV in the barrel pseudorapidity region ($|\eta| < 3$) [3].

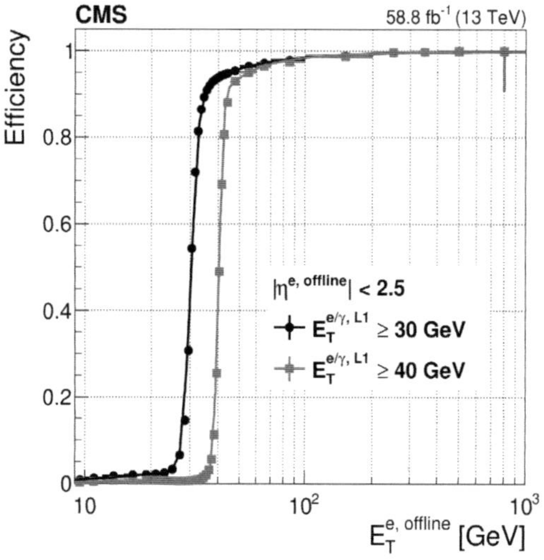

Figure 3. The L1 e/γ trigger efficiency as a function of the offline reconstructed electron E_T for thresholds of 30 and 40 GeV[3].

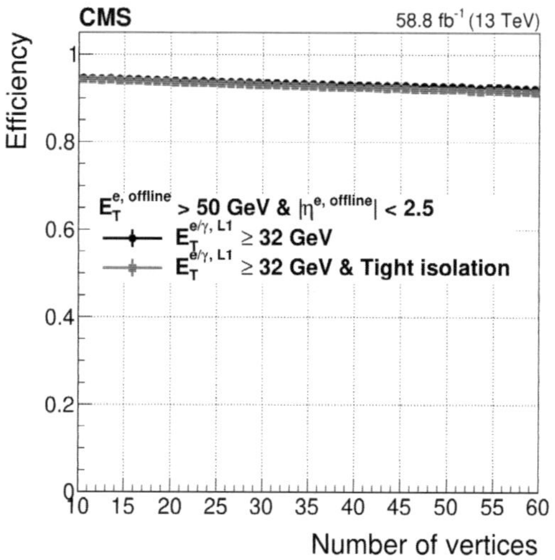

Figure 4. The L1 e/γ isolated trigger efficiency, presented as a function of the offline reconstructed vertices[3].

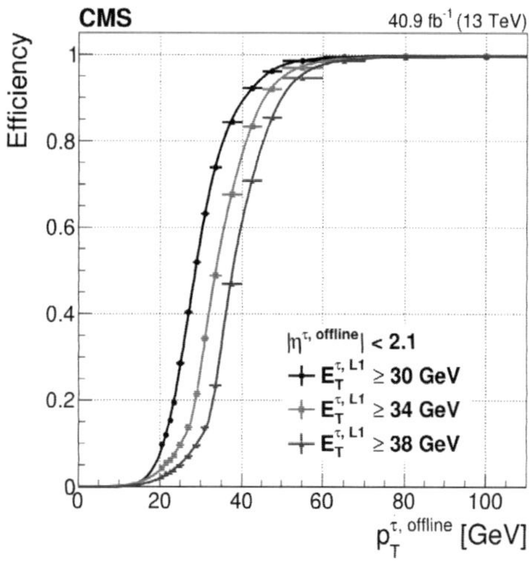

Figure 5. The L1 τ trigger efficiency, as a function of the offline reconstructed τ lepton p_T, for typical thresholds of 30, 34, and 38 GeV [3].

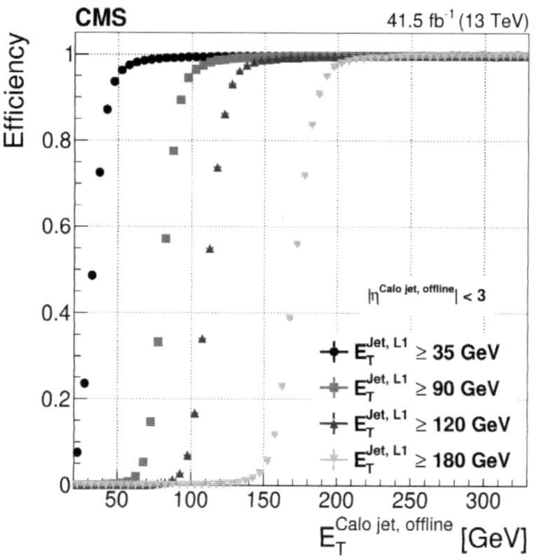

Figure 6. Efficiency curves for the Level-1 jet trigger for the barrel pseudorapidity range [3].

4. Performance of the High Level Trigger System

The High Level Trigger system reduces the L1 Trigger output rate of about 100 kHz further down to around 1 kHz at the end of Run 2, with an estimated HLT rate at around 1.5 kHz for the Run 3. The HLT uses offline reconstruction algorithms, but optimised to run around hundred times faster. It is composed of hundreds of HLT paths, targeting a broad range of event topologies, that consist of reconstruction and filtering modules executed in sequences. When an event is rejected by a filter, the subsequent modules within the same path are not run.

Figure 7 presents the tracking efficiency as a function of the simulated track p_T for the Run-2 HLT tracking, given in blue, and the Run-3 HLT single-iteration tracking, shown in red [4]. The Run-3 tracking clearly improved over the previously used for the Run-2. The efficiency at large track p_T is reduced due to the presence of a large number of tracks in the core of high p_T jets. Figure 8 shows the tracking fake rate as a function of the reconstructed track p_T. In the Run-3 HLT tracking, unlike in Run-2, no track with $p_T < 0.3$ GeV is reconstructed [4]. As a consequence, the total amount of fake tracks in Run-2 HLT tracking is sensibly larger with respect to Run-3.

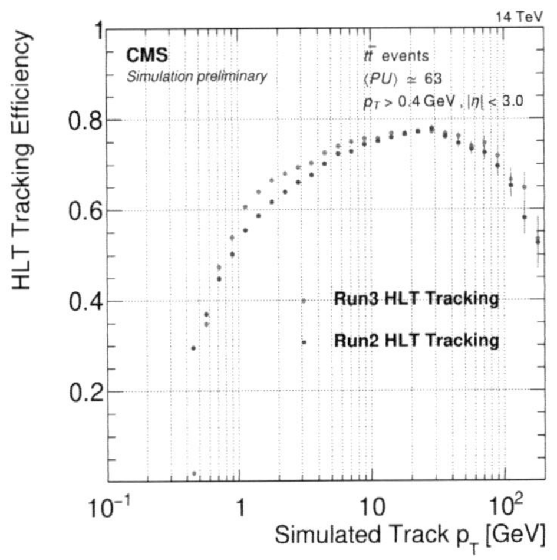

Figure 7. The tracking efficiency is shown as a function of the simulated track p_T for the Run-2 HLT tracking (blue) and the Run-3 HLT single-iteration tracking (red)[4].

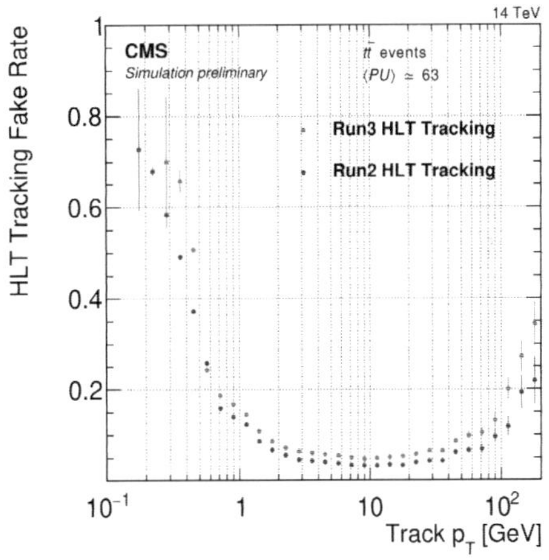

Figure 8. The tracking fake rate is shown as a function of the reconstructed track p_T for the Run-2 HLT tracking (blue) and the Run-3 HLT single-iteration tracking (red)[4].

Figure 9 shows the efficiency to identify b-jets online, compared to the corresponding offline reconstruction [5]. A neural network based classifier called Deep CSV (Combined Secondary Vertex) is used at CMS since 2017 for the identification of b-tagged jets. The improvement over the previously used CSV algorithms ranges from 5 to 15 % at the same light flavor efficiency. The distribution of the Deep CSV discriminator for online PF jets is shown in Figure 10, where different colours present the individual contributions that are coming from different jet flavors.

The performance of the online selection of the electrons at the High Level Trigger is shown in Figure 11, for one characteristic HLT path, with respect to an offline reconstructed electron as a function of the electron p_T, obtained for different η regions using the full 2018 dataset [6]. The efficiency of a characteristic HLT path that requires a photon with the p_T higher than 200 GeV, used for example in Supersymmetry and other new physics searches, is in Figure 12 [7].

The efficiency to reconstruct jets at the HLT is measured in single muon events and shown in Figure 13 as a function of the offline reconstructed jets [8]. Both offline and online jets are reconstructed using the PF algorithm. The HLT jets were matched with the offline jets. The absolute value of the pseudorapidity for the jets is restricted to 2.4. The efficiency of high-level

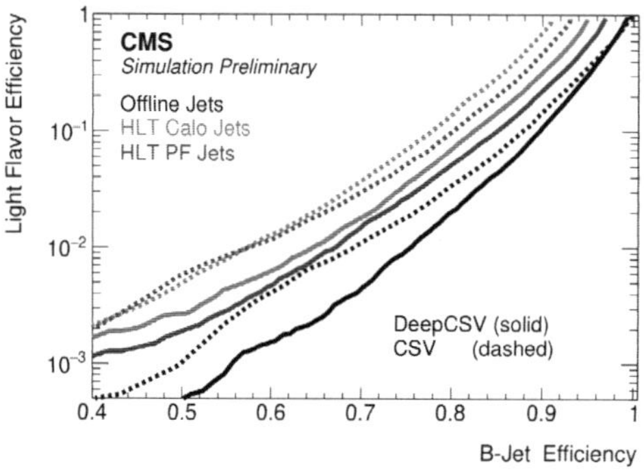

Figure 9. Performance of the online (red and blue) and offline (black) b-jet identification efficiency demonstrating the probability for a light-flavor jet to be misidentified as b-jet as a function of the efficiency to correctly identify a b-jet[5].

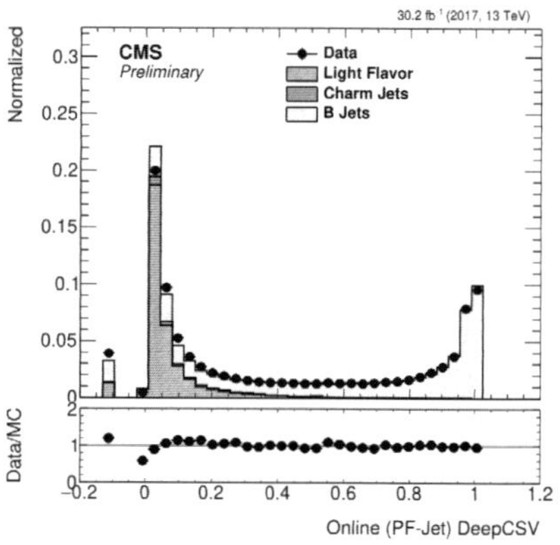

Figure 10. The DeepCSV discriminator distribution for online (PF-Jets). Different colours show the contributions in simulations from different jet flavors[5].

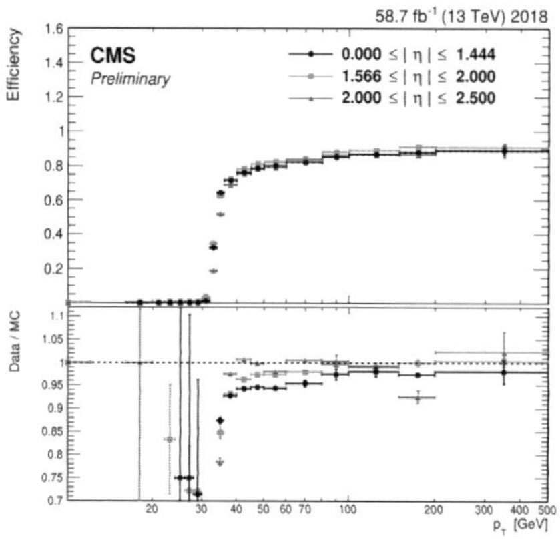

Figure 11. The efficiency of a single electron trigger with p_T >32 GeV, with respect to an offline reconstructed electron as a function of the electron p_T, obtained for different ηregions using the full 2018 dataset [6].

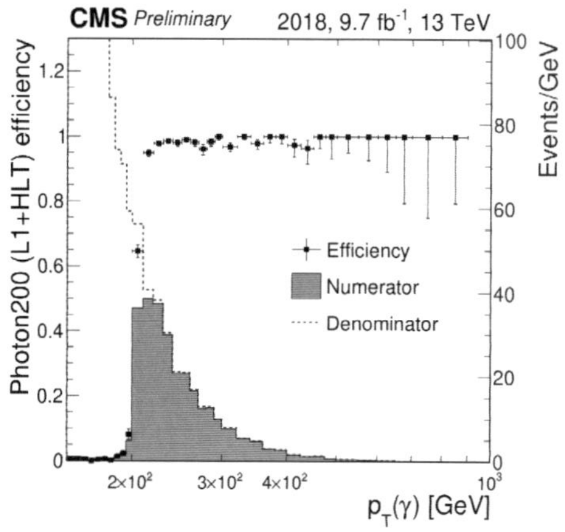

Figure 12. The efficiency of a High-Level Trigger (HLT) algorithm that requires a photon with a transverse momentum p_T greater than 200 GeV[6].

trigger requiring isolated single muon with p_T >24 GeV is shown in Figure 14 as a function of muon p_T [9], before and after the updated reconstruction in 2018. The efficiency is estimated with respect to the offline muon matched to L1 trigger object with $\Delta R < 0.3$ as well as passing tight ID and PF based isolation requirements.

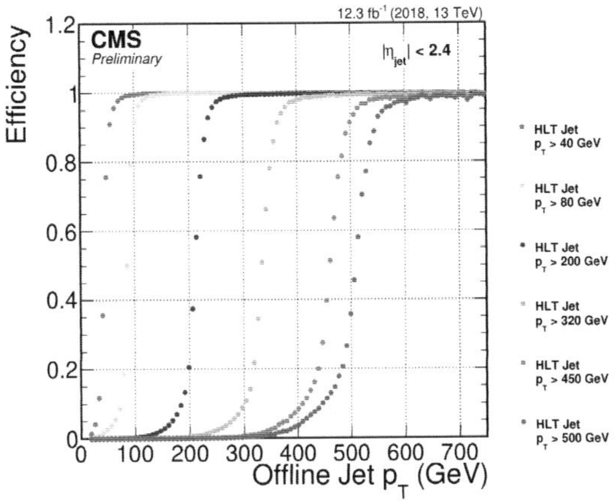

Figure 13. The efficiency to reconstruct jets at the HLT as a function of the offline reconstructed jet p_T[8].

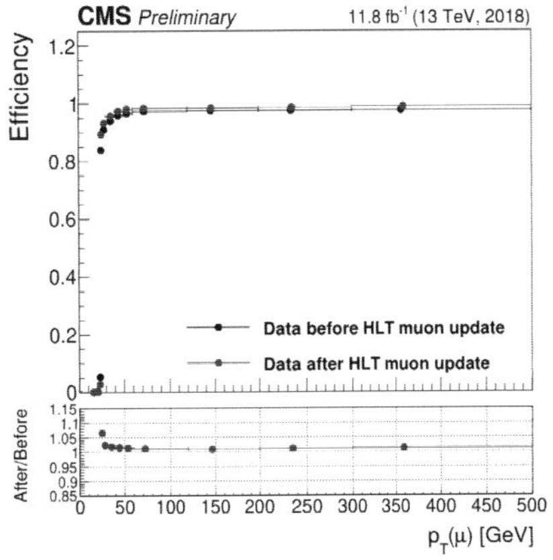

Figure 14. The efficiency of high-level trigger requiring isolated single muon with p_T >24 GeV as a function of muon p_T [9].

5. Trigger Menus for Data Selection

The trigger menu represents a large set of selection criteria that is imposed during the online event selection, enabling to fulfil the broad physics program of the CMS experiment. Each trigger menu consists of general or multi-purpose triggers, but also some very specific triggers used only in particular studies and searches, as well as backup triggers. There are separate trigger menus used at the L1 and at the HLT, with about 400 and 600 items, respectively. Figure 15 presents fractions of the total L1 rate (100 kHz) allocation for single- and multi-object triggers and cross triggers in a typical CMS physics menu during Run 2 [3]. The dominant fraction of the total L1 rate is taken by the single and multi objects, with less rate given to the cross triggers.

Figure 16 shows the HLT rate per physics group during the data taking in 2018, presenting the total, shared and pure rates [10]. The total rate is the rate from an event assigned to all groups that trigger the event. The shared rate is the rate from an event shared equally among all groups that trigger the event, while the pure rate is the rate from an event assigned to a given group if it is the only one that triggers the event. The rates were evaluated by running the HLT menu on a commissioning data set. The latter consists in a fraction of events passing the Level-1 trigger, without any further HLT requirement. The average recorded instantaneous luminosity in the input data was $1.2 \times 10^{34} cm^{-2}s^{-1}$. All rates were scaled to $2.0 \times 10^{34} cm^{-2}s^{-1}$.

6. GPU-based acceleration at High Level Trigger

At CMS experiment, a heterogeneous architecture will be used in the online reconstruction in the Run 3, comprising CPUs and GPUs [11]. The pixel and pixel-based tracking, as well as the ECAL and HCAL local reconstruction have already been ported to GPUs and more reconstruction code is foreseen to be ported to GPUs in the near future. The pie-charts in Figures 17 and 18 represent the distribution of CPU and GPU time, respectively, spent in different instances of CMSSW modules (outermost ring), their corresponding C++ class (one level inner) and grouped by physics object or detector (innermost ring), while the empty slice indicates the time spent outside of the individual algorithms. The timing is measured on pileup 50 events from 2018, running 4 jobs in parallel, with 32 threads each, on a node (2x AMD "Rome" 7502) with SMT enabled, and, in case of heterogeneous architecture, additionally equipped with

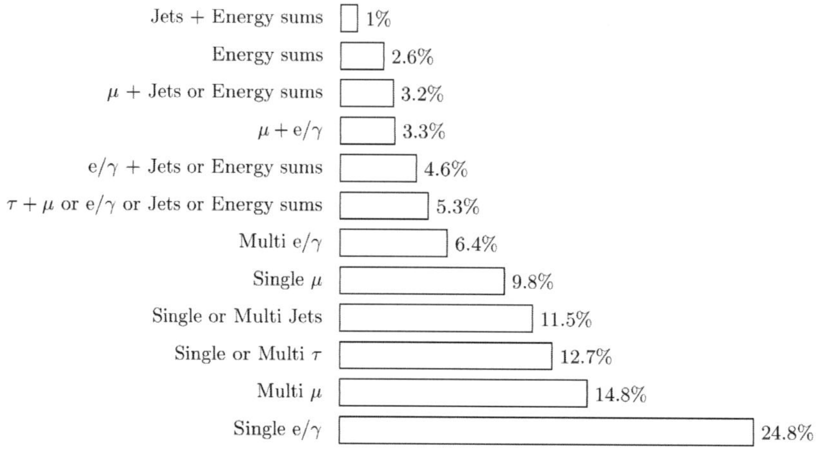

Figure 15. Fractions of the 100 kHz rate allocation for single- and multi-object triggers and cross triggers in a typical CMS physics menu during Run 2 [3].

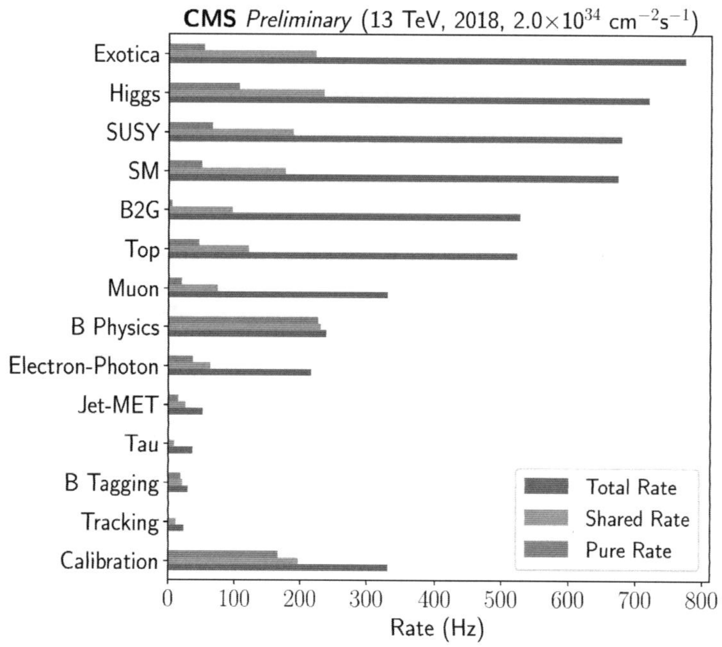

Figure 16. The HLT rates consumed by each CMS physics group during 2018 data taking [10].

NVIDIA T4 GPU. Around 25% of the CPU time is offloaded to GPU at present.

Journal of Physics: Conference Series **2375** (2022) 012003 doi:10.1088/1742-6596/2375/1/012003

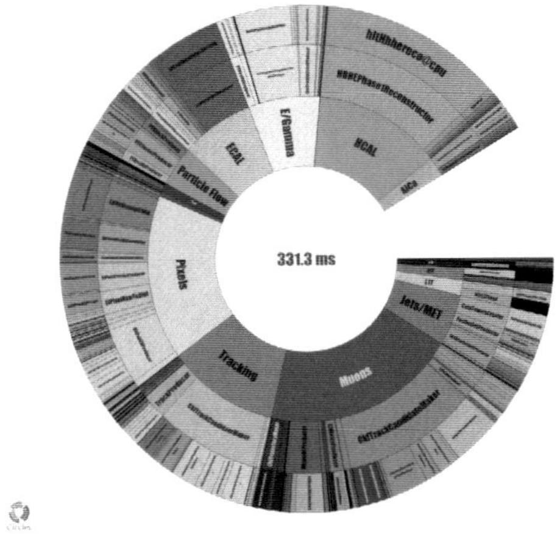

Figure 17. The pie-chart shows the distribution of CPU time in different instances of CMSSW modules [11].

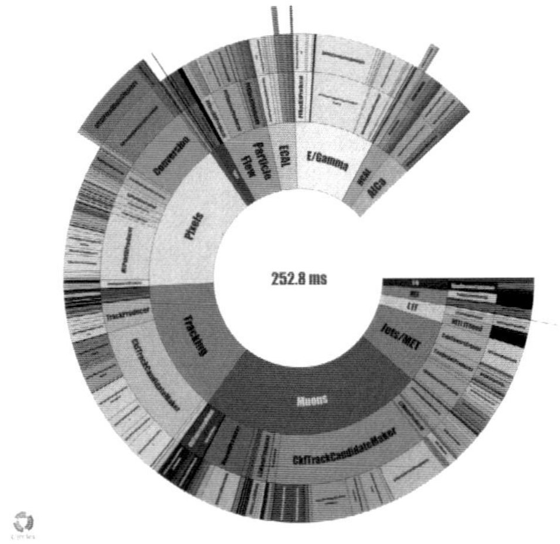

Figure 18. The pie-chart shows the distribution of CPU and GPU time in different instances of CMSSW modules [11].

7. Data Scouting and Data Parking

Alternative strategies exist with respect to the standard online data taking, as described above, in order to surpass the limited storage and computer processing resources. One option is referred as Data Scouting, where only a small summary of the reconstructed event quantities is saved and not all the raw data, thus reducing the event size, in order to be able to record the events with higher rate [12]. An example of the usage of Data Scouting is presented in Figures 19 where the di-muon events are recorded largely unconstrained by requirements on muon kinematics otherwise imposed by CMS data acquisition and event reconstruction workflows. For low invariant mass dimuon events the data taking efficiency is improved by one or two orders of magnitude. Another option is Data Parking, where the data is parked on tape, skipping the prompt reconstruction, thus reducing the required computing resources, and reconstructing the data later, in shutdown period when the experiment is not taking data [13].

8. Summary and Outlook for Run 3

The CMS Trigger System has manifested its robustness and flexibility and has proven during the Run 1 and Run 2 to be able to deal with a large number of events to fulfil the CMS physics goals. Excellent performance was obtained in Run 2, from sharp efficiency curves to only a moderate pileup dependence. During the course of Run 2 and in the Long ShutDown 2 (LS2), many new technologies were integrated in the CMS Trigger System, and many trigger algorithms were improved and innovated [14]. The Phase-I L1 trigger upgrade had occurred in 2016, introducing finer calorimeter granularity, resulting in improved energy and position resolution, while remaining within the rate constraints. The pileup subtraction algorithms were implemented in the L1 trigger during the Run 2, at a large benefit of the L1 performance [3]. The High Level Trigger reconstruction was improved following the Phase-I Pixel upgrade [15] from 2016 to 2017. HCAL had the Phase-I Upgrade in endcap and barrel, during 2018 and 2019, respectively [16].

A particular improvement is introduced for the Run 3, referred to as the heterogeneous

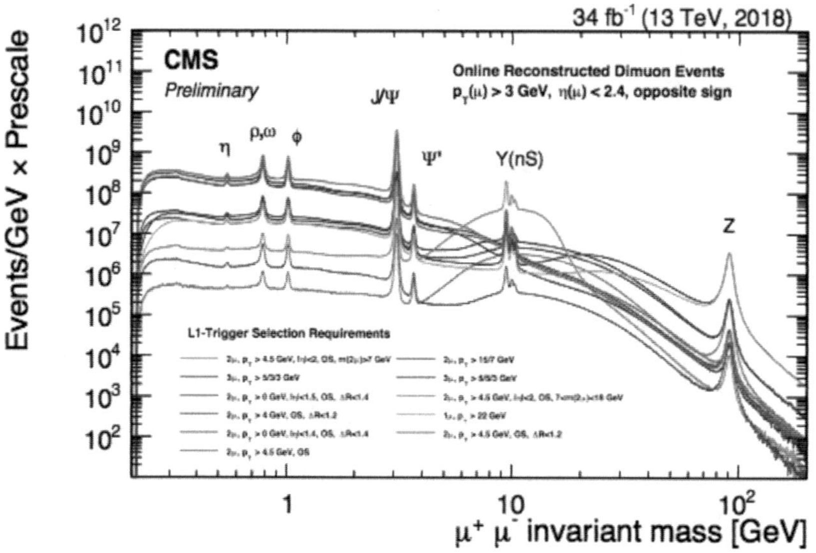

Figure 19. Dimuon invariant mass spectra reconstructed in the High Level Trigger system of the CMS detector for various muon Level-1 trigger requirements deployed by the CMS collaboration in 2017. For a subset of Level-1 requirements CMS recorded only a fraction of the data [12].

reconstruction comprising CPU and GPU. In the Run 3, it is planned to expand the reach to high rates and more exotic phase spaces, like developing special triggers for long-lived particles [17].

References

[1] The CMS Collaboration, JINST 3 (2008) S08004.
[2] The CMS Collaboration, JINST 12 (2017) P10003.
[3] The CMS Collaboration, JINST 15 (2020) P10017.
[4] The CMS Collaboration, CMS-DP-2022/014.
[5] The CMS Collaboration, CMS-DP-2019/042.
[6] The CMS Collaboration, CMS-DP-2020/016.
[7] The CMS Collaboration, CMS-DP-2018/049.
[8] The CMS Collaboration, CMS-DP-2018/037.
[9] The CMS Collaboration, CMS-DP-2018/034.
[10] The CMS Collaboration, CMS-DP-2018/057.
[11] The CMS Collaboration, CMS-DP-2021/013.
[12] The CMS Collaboration, CMS-DP-2018/055.
[13] The CMS Collaboration, CMS-DP-2019/043.
[14] https://home.cern/press/2022/CMS-upgrades-LS2.
[15] The CMS Collaboration, CERN-LHCC-2013-011; CMS-TDR-12.
[16] The CMS Collaboration, CERN-LHCC-2012-016 ; CMS-TDR-11.
[17] The CMS Collaboration, Phys. Rev. Lett. 127 (2021) 261804.

Performance studies of Micromegas electronics in a high radiation environment at the CERN Gamma Irradiation Facility (GIF++)

Foteini Kolitsi[a], Daniel Bita[a], Ioannis Mesolongitis[a], Chara Kitsaki[b], Theodoros Alexopoulos[b], Efstathios Kyriakis-Bitzaros[a], Katerina Zachariadou[a]

[a] University of West Attica, Athens, Greece, [b] National Technical University of Athens, Greece

E-mail: foteini.kolitsi@cern.ch

Abstract. The ATLAS experiment at CERN undergoes a series of upgrades, in line with the upgrades (Phase-I and Phase-II) of the Large Hadron Collider (LHC), to cope with the steadily increasing instantaneous luminosity that ultimately is expected to reach 7.5×10^{34}cm^{-2} s^{-1} at the High Luminosity LHC (HL-LHC) era. The most challenging upgrade project of the ATLAS experiment concerns its Muon Spectrometer's inner wheel-shaped detection stations (Small Wheels) located close to and on either side of the pp beams' interaction point. Another type of detection system (New Small Wheels) replaced the two Small Wheels so that the ATLAS maintains its performance while preserving the acceptance of critical physics signatures under the harsh conditions of higher data and radiation rates (up to 25kHz/cm^2 at the HL-LHC accelerator). The NSWs are a complex detection system (each wheel comprises 2.4x10^6 readout channels)that employ two novel micro-pattern gaseous detector technologies, the Micromesh Gaseous Structure (Micromegas, MM) and the small strip Thin Gap Chambers (sTGCs), mainly to improve the trigger performance and to provide precise spatial measurements. New electronic boards residing on the detectors were designed and manufactured, hosting high radiation and magnetic field tolerant custom-made ASICs.This paper presents extensive tests carried out at the new CERN Gamma Irradiation Facility (GIF++) by combining a muon beam (of 12kHz) with a ^{137}Cesium radiation source of adjustable rate (up to ten times the expected rate at the HL-LHC conditions), to assess the performance of the Micromegas chambers and their readout electronics. Data from NSW Micromegas modules of different detection areas, having different positions on the wheels, operating with different high voltage and filling gas have been read out and analyzed by placing the electronics on PCBs located at various distances from the pp beam pipe. The spatial resolution of the precision coordinate is found to be well below the required 100μm, and the excellent alignment of the precision layers has been verified. Results confirm that all types of Micromegas modules tested have a linear response up to a rate around four times the expected at HL-LHC.

Keywords: Micromegas, Resistive-strip, New Small Wheel electronics, MicroMegas Performance, New Small Wheel, Muon spectrometer upgrade, ATLAS experiment upgrades, ATLAS experiment

1. Introduction
The Large Hadron Collider (LHC) [1] at CERN to extend its physics potential is scheduled to increase its instantaneous luminosity. Its first upgrade (Phase-I) was accomplished during the long technical shutdown (2019-2022), and during its current operation (run3), an instantaneous luminosity of 2-3$\times10^{34}$cm^{-2} s^{-1}, is expected to reach. After its next (Phase-II) upgrade

(2025-2027), the High Luminosity (HL-LHC) will begin its operation with a foreseen maximum instantaneous luminosity of $7\times10^{34}\text{cm}^{-2}\ \text{s}^{-1}$.

Due to the increased LHC instantaneous luminosity, the higher data and trigger rates and the higher beam-induced background would degrade the ATLAS [2] subdetectors' efficiency and raise the fake muon triggers, especially in regions closer to the beam pipe.Therefore, to keep its excellent performance in terms of efficiency, resolution, and background rejection while preserving the acceptance of critical signatures for many physics channels, the ATLAS experiment has to upgrade its detectors and electronics.

The challenge of the ATLAS upgrades is its Trigger [3] and Data AcQuisition [4] system (TDAQ) and trigger algorithms to maintain low transverse momentum trigger thresholds at manageable trigger rates in a much higher background environment. For this, the forward regions (end-caps) of the ATLAS Muon Spectrometer [5] are of particular importance.

The Muon Spectrometer is arranged in three coaxial cylindrical detector systems around the beam axis (barrel region) and four parallel plates on both sides of the barrel, perpendicular to the beam axis (end-caps). The expected hit rate at the innermost end-caps (the "Small Wheels") is expected to be up to 15 kHz/cm^2, necessitating, therefore, its replacement with another type of detection system, named "New Small Wheels" (NSWs) [6].

Each NSW accommodates two novel detector technologies; the small-strip Thin Gap Chambers (sTGC)[7], mainly for triggering, and the Micro-Mesh Gaseous Structure (MicroMegas, MM), mainly for the precise measurement of the muon's position and momentum [8].

This work aimed to investigate the performance of the NSW's MM detectors in a high radiation of up to ten times the rate expected at the HL-LHC conditions (25kHz/cm^2).

The MM operation principle and the NSW layout are briefly described in sections 2 and 3. Then, section 4 describes the MM trigger and readout electronics used for the tests. Section 5 presents the hardware testing setup and the tests performed. Finally, section 6 summarizes the results of the tests.

2. The Micromegas Detector

The MM detectors (figure 1) are gaseous detectors that consist of three planes, the cathode (drift), the anode (readout), and a thin mesh plane (micromesh). The micromesh divides the gas volume into two regions; the conversion/drift gap (5 mm) and the amplification gap (128 µm). Charged particles entering the conversion/drift region ionize the gas and create electron-ion pairs. The produced ions drift towards the cathode while the electrons drift towards the micromesh. Because of the much higher electric field of the amplification region, the created electrons pass through the mesh to the amplification region, where electron avalanches occur. The induced charge on the readout strips forms a fast pulse read out by the front-end electronics. The resistive strips, separated from the readout strips by an insulating layer, serve to make the chamber discharge tolerant [9]. Both the anode and cathode are based on printed circuit boards (PCB)[10].

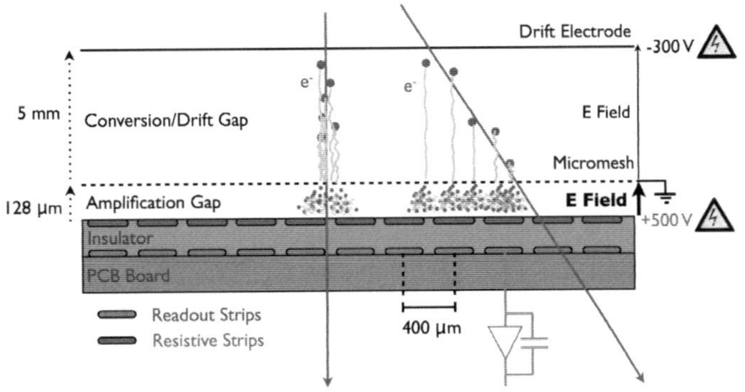

Figure 1: Operation principle of a resistive MicroMegas detector[10]

3. The New Small Wheel project

The NSW system [6] is composed of two wheels, each one placed at the corresponding end-cap (A or C) of the ATLAS Muon spectrometer, and employs MM and sTGC detectors. Each NSW wheel comprises eight large and eight small sectors, trapezoidal shaped and fitted in a metal circle structure, with a diameter of 10 meters (figure 2a). Each sector comprises two MM wedges mounted on both sides of a metallic central spacer frame and sandwiched by two sTGC wedges (figure 2b). Each wedge of a sector is assembled radially from two separate trapezoidal shape quadruplet modules (figure 2c), trapezoidal shaped and off different sizes, each consisting of four detection layers (quadruplet). The SM1 (LM1) module of the Small (Large) sector having a smaller radius (closer to the beampipe), whereas the SM2 (LM2) is the one with the higher radius. The chambers' areas of the small (SM1, SM2) and the large (LM1 and LM2) modules are 2m^2 and 3m^2, respectively.

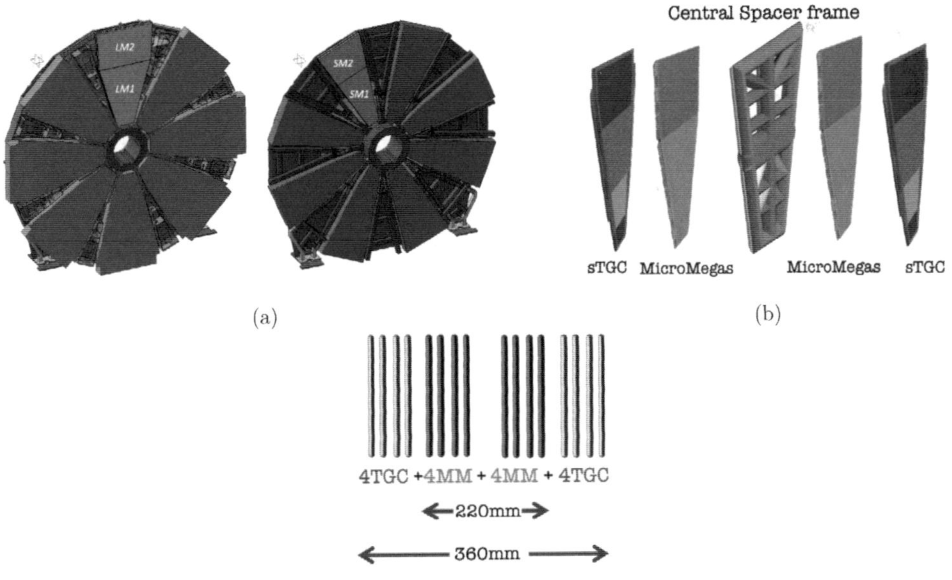

(a) (b) (c)

Figure 2: The NSWs layout: a) the large and the small NSW sectors b) View of a sector with two sTGC and two MM quadruplets on both sides of the spacer frame c) dimensions of a wedge

The MM quadruplets are composed of two readout panels (named eta and stereo) that carry the readout boards and three drift panels comprising the drift electrode and the micro-mesh, constructed as described in [11]. Each of the two readout panels has on each side two detection layers. On the eta readout panel, the readout strips of the layers (named η Layer 1 and 2) are oriented parallel to the trapezoid bases for measuring the η (precision) coordinate. In contrast, on the stereo panel, the strips of the layers (named stereo Layer 3 and 4) are tilted ($\pm 1.5°$) relative to the η-strips for measuring the azimuthal coordinate. Each readout layer in the SM1 and LM1 modules is segmented into five readout boards (PCBs numbered 1,2,3,4,5), while in the SM2 and LM2 modules, each readout layer has three readout boards (PCBs numbered 6,7,8), as illustrated in figure 3. MM chambers with different detection areas and positions relative to the beam pipe (an SM1 and an LM2 module) and with their readout electronics placed on all the detection layers of selected PCBs have been tested to investigate their performance, as described in section 5.2.

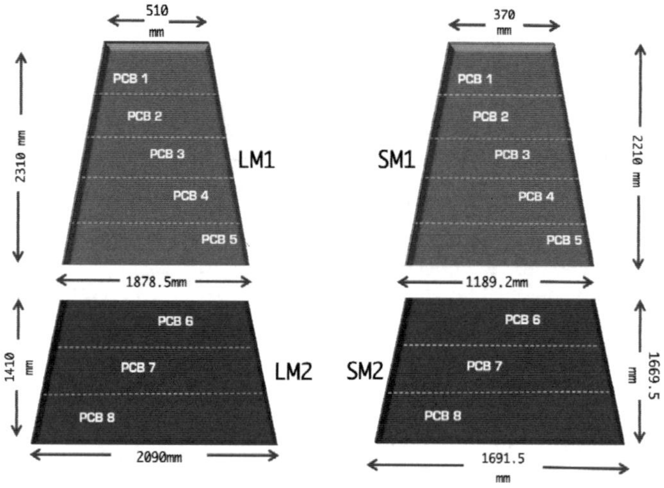

Figure 3: PCB mapping

4. The Micromegas Electronics

The NSW electronics (figure 4) are classified into two categories; on-detector and off-detector electronics. The on-detector electronics reside on the detectors and are inside the detector area (cavern), whereas the off-detector are located in the USA area (a few meters outside the cavern). The Micromegas' readout electronics used for the tests are shown in the lower part of figure 4. These are a) the front-end boards [12] and b) the Level-1 Data Driver Card (L1DDC) [13].

The front end boards host a) eight Venetios Micromegas (VMM3) [14] that provide the peak amplitude and the time relative to the bunch crossing clock or other trigger signals via 64 channels, where each channel is connected to a readout strip, (b) one Read Out Controller (ROC) [15] that transmits the data to the readout path and (c) one Slow Control Adapter (SCA) [16] used for the configuration and monitoring of all the on-detector boards.

The main radiation tolerant ASICs that the L1DDC hosts are the GigaBit Transceiver (GBTX)[17], a high-speed serializer/deserializer ASIC responsible for the bidirectional data transmission between the front-ends and the off-detector network interface system (Front-End LInk eXchange, FELIX)[18].

Data from multiple VMM3s are collected by the ROC ASIC, which transmits them through serial links at 320Mb/s to the GBTX ASIC of the L1DDC board. Then, the GBTX transmits them via a single fiber to FELIX and finally to the Read Out Devices (ROD) [19] and the Detector Control System (DCS)[20].

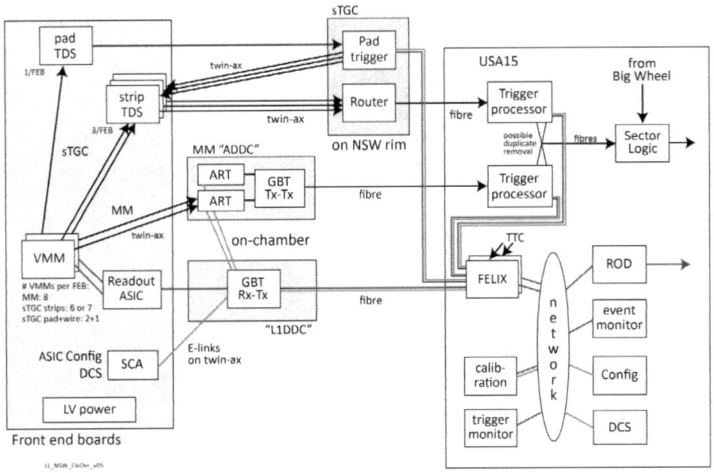

Figure 4: The NSW electronics: trigger path of the sTGCdetecors (up), trigger path of the MM detectors (middle), and data acquisition path of both the MM and the sTGC detectors (down).

5. The Micromegas tests at the GIF++

Extensive tests have been carried out in the new CERN Gamma Irradiation Facility (GIF++) [21] located in the North Area of the Super Proton Synchrotron (SPS). The GIF++ combines a high-energy particle beam (80-150 GeV) in the SPS H4 beam line with a 14 TBq ^{137}Cesium source of 662keV to simulate the radiation intensity expected at the HL-LHC. A system of absorbers can tune the flux of the emitted gamma rays.

5.1. The Experimental Setup

The experimental setup in the SPS H4 Experimental Hall (figure 5) consists of the MM modules, two scintillators (SC1, SC2) to provide the trigger, a 14 TBq Caesium (^{137}Cs) gamma source, and an 80-150 GeV muons beam with a rate of 12kHz/cm^2 rate). Three modules of different types (SM1 and LM2), operated with isobutane gas and a small prototype (Tz) with an active area of 10 x 10 cm^2 were tested. Tz uses one layer, and the readout is provided by one board connected to 256 strips. Due to its small capacitance, it has an ideal linearity behavior and was used for reference. The modules were positioned sequentially and vertical with respect to the beam, between the scintillators. The distances with each other are shown are shown in figure 5.

The irradiator was placed at about 1m distance from the SM1 module (corresponding to a gamma rate of (250kHz/cm^2), and its rate was attenuated down to the rate expected at HL-LHC 25 kHz/cm^2 by using a set of absorbers. For the SM1 module, the MMFE8 boards were placed on PCBs 2 and 3, whereas for the LM2, on PCBs 6 and 7.

In total, thirty-three MMFE8s were used (sixteen in each module and one in the Tz chamber), and four L1DDCs for obtaining the Level-1 accept (L1) data. The muon beam was crossing the whole Tz chamber and the PCB 2 and 6 of the SM1 and LM2 modules, respectively.

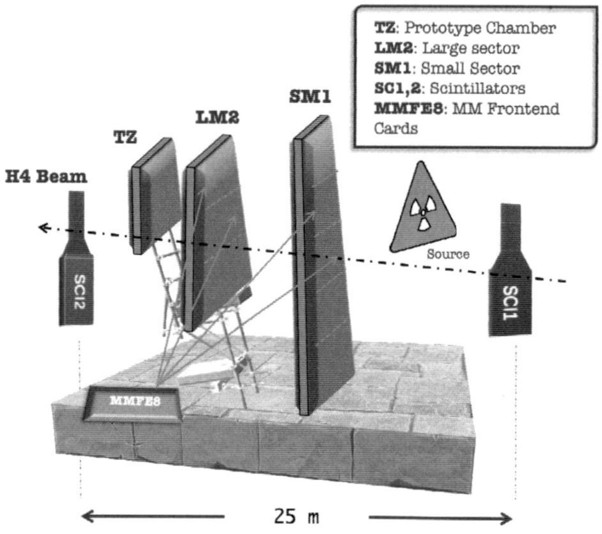

Figure 5: The experimental setup at GIF++.

5.2. Results

Tests under all source and muon beam operation combinations have been performed, and measurements under different values for the high voltage and the VMM3s' parameters have been taken to investigate the modules' performance.

The following paragraphs present the results of the measurements taken with the VMM3s'gain set to 9mV/fC and the integration time of the pulses equal to 200ns. All chambers were filled with isobutane gas mixture and operated at a high voltage of 520V.

To evaluate the spatial resolution of the precision coordinate, the modules were irradiated with the muon beam (12kHz/cm^2 rate) in the absence of the ^{137}Cs source. Their readout planes were perpendicular to the beam. Figure 6 shows the beam profile reconstructed using stereo layers of the SM1 and LM2 modules. A passivation area at the SM1 module can be noticed.

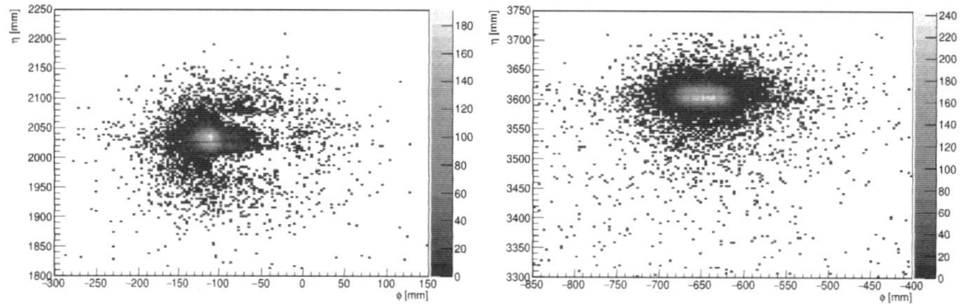

Figure 6: The beam profile in the η-ϕ view, detected by the SM1 (left) and LM2 (right).

For the evaluation of the spatial resolution of the precision coordinate, the cluster position on two precision layers has been reconstructed using the charge centroid method, as described in [22]. Assuming that the spatial resolutions of the two layers are equal, the combined spatial resolution can be estimated by the difference in the cluster position on the two layers. The distribution of the difference was fitted with a Gaussian function figure 7, and the derived standard deviation (σ) was divided by $\sqrt{2}$ [23]. The core distribution of the difference (95%

of the events analyzed) that gives an estimation of the intrinsic detector resolution, was fitted with a single Gaussian (represented by a black line in figure 7), while a double Gaussian fit was applied to account for the tails (red lines in figure 7). A resolution of 64µm (84 µm) for the SM1 (LM2) module has been obtained, below the 100µm required for the precision coordinate. It can also be noticed that the mean value of the position distributions (for both SM1 and LM2 modules) is almost equal to zero, indicating a good alignment of the precision layers. (For the LM2, the observed offset is about 50µm, and can be considered negligible compared to the 450µm pitch of the strips).

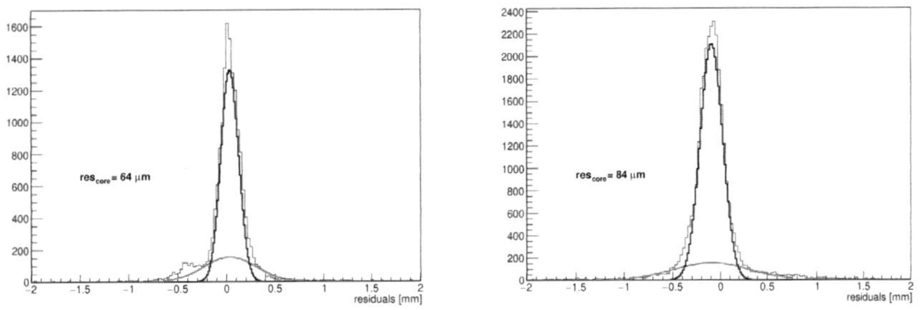

Figure 7: The spatial resolution of the coordinate for perpendicular tracks on the SM1 (left) and the LM2 (right) module.

The response of the detectors under high radiation rates has been evaluated using the muons beam and the ^{137}Cs source simultaneously. As a metric, the hits occupancy as a function of the source rate has been studied.

The theoretical values expected for the hit occupancy (HOth) are calculated using the formula:

$$HO_{th} = Rate \times sectorPitch \times PCBLength \times NumberOfHitsPerCluster \times ReadOutWindow \quad (1)$$

where Rate is the radiation rate of the ^{137}Cesium source (in kHz/cm^2), sectorPitch is the gap between two consecutive strips, PCBLength is the length of the PCBs, NumberOfHitsPerCluster is the Number of Hits that form a cluster, and ReadOutWindow is the duration of the integration of a pulse performed by the VMM3. According to the previous formula, the dependence of hit occupancy versus the rate is linear (figure 8). Given that the SectorPitch is 0.0425cm (0.045cm) for the SM1 (LM2), the PCBLength is 68cm, the NumberOfHitsPerCluster equals 5, and the ReadOutWindow is 200ns, the expected hit occupancy at the HL-LHC rate (20 kHz/cm^2) is 5.8%.

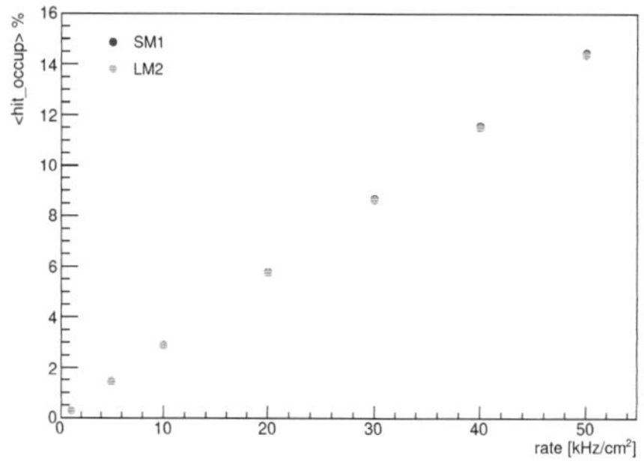

Figure 8: The theoretical values expected for the hit occupancy (%) of SM1 and LM2 modules as a function of the radiation rate. (The difference of the hit occupancy values of SM1 and LM2 are too small to be visible).

For the hits occupancy study, the maximum source's intensity (250kHz/cm^2) at a distance of 1m from the modules was attenuated down to the expected at the HL-HC rate using a set of absorbers and was calculated as:

$$HO_{exp} = \frac{Average\ Number\ Of\ Strip\ Hits}{Number\ Of\ Strips} \tag{2}$$

where the Average Number of Strip Hits was measured by analyzing the Level-1 data received during a particular test.

Figure 9 shows the hit occupancy as a function of the source's inverse attenuation for the SM1 and LM2 modules and the prototype Tz. As seen, the SM1 and LM2 modules saturate approximately at a rate of 25% of the maximum source's intensity which corresponds to four times the rate expected at HL-LHC.

The saturation of the modules at high rates is due to the faster gas ionization, consequently in higher electron production in the chambers' conversion region. The charge of the strips (because of their long length) cannot evacuate fast enough, leading to a significant number of strip hits in a given time interval. Since the VMM3s can read only one hit in that interval, any other hit during that time window is missing.Indeed, as shown in figure 9c, the Tz chamber has a linear behavior since its strips have a shorter length.

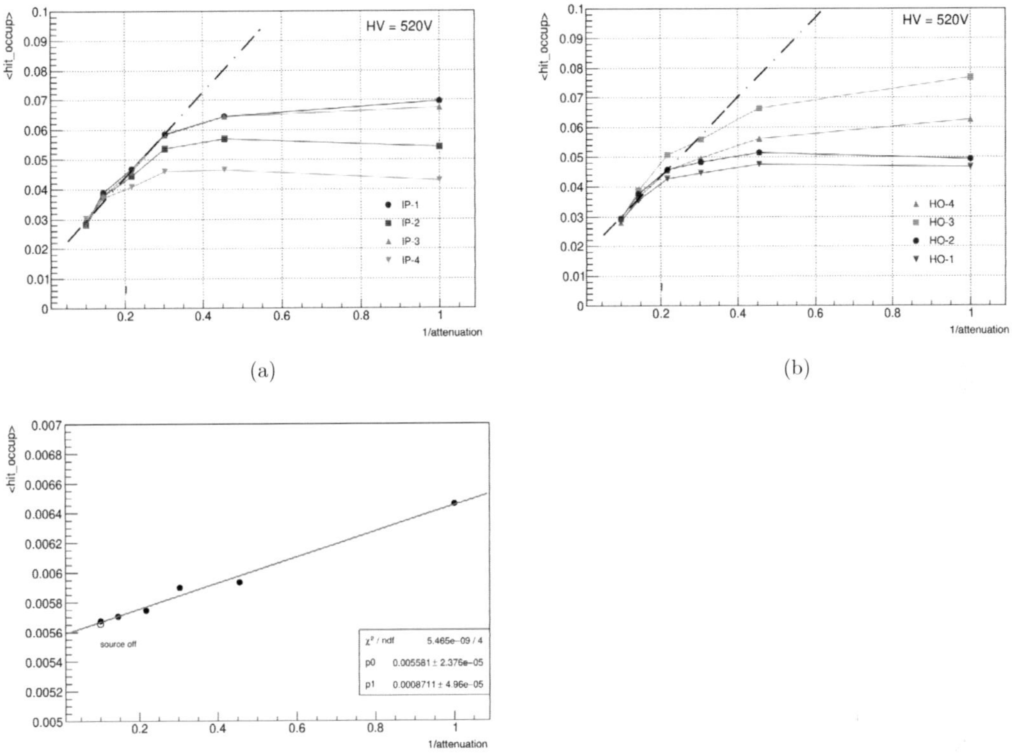

(a) (b) (c)

Figure 9: Hit occupancy versus the inverse attenuation of the source's intensity for all four detection layers of (a) the SM1 module (IP-1 to IP-4), (b) the LM2 module (HO-1 to HO-4), and (c) the Tz chamber.

The impact of the high voltage on the performance of the detectors has also been studied. In the tests, the muon beam was off, and a random trigger of approximately 500 Hz was used. Figure 10 shows the hit occupancy for the SM1 module at different high voltages. It can be noticed that at low high voltages, the hits' linearity is preserved up to high radiation rates. On the contrary, at high voltages, the saturation appears at lower rates due to the higher amplification in the amplification gap leading to an overlap between the signals that the VMM3s cannot distinguish; consequently, the number of hits read by the VMM3s decreases.

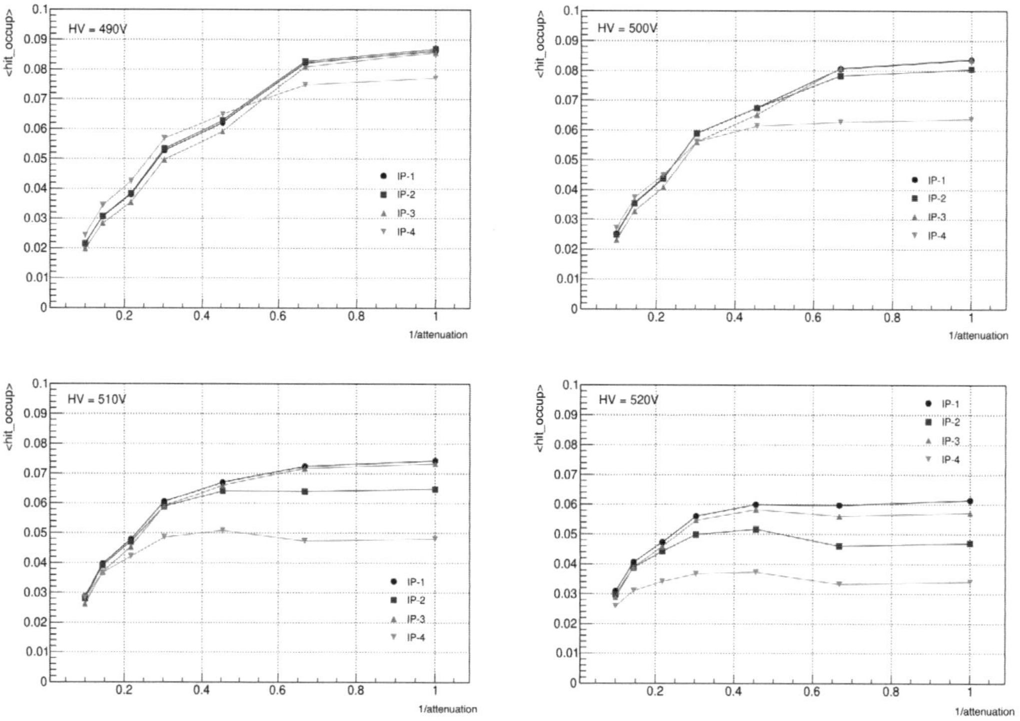

Figure 10: Hit occupancy, for the SM1 module, as a function of the radiation rate using different high voltage values.

6. Conclusion

The performance of two Micromegas modules (SM1 and LM2) and their readout electronics have been tested at the CERN GIF++ facility using a high energy muon beam of 12kHz/cm^2 rate and a ^{137}Cesium source of rates up to ten times the rate expected in the forthcoming HL-LHC. For tracks perpendicular to the modules' planes, a spatial resolution of 64µm (84 µm) for the SM1 (LM2) module has been obtained, well below the 100µm required for the precision coordinate. Extensive tests performed in the presence of both the muon beam and the irradiator indicate that the hit occupancy preserves its linearity up to a rate of about four times the rate expected at HL-LHC.

The saturation of the electronics without the ^{137}Cs irradiator was also studied for different high voltage values. As expected, the high voltage is a trade-off parameter between high values, essential for the efficient ionization of the gas, and low ones, so that the electronics' saturation occurs at higher rates.

7. Acknowledgments

We acknowledge support of this work by the project "DeTAnet: Detector Development and Technologies for High Energy Physics" (MIS 5029538) which is implemented under the action "Reinforcement of the Research and Innovation Infrastructure" funded by the Operational Programme "Competitiveness, Entrepreneurship and Innovation" (NSRF 2014–2020) and co-financed by Greece and the European Union (European Regional Development Fund). The research work was supported by the Hellenic Foundation for Research and Innovation (HFRI) under the HFRI PhD Fellowship grant (Fellowship Number: 800)

8. References

[1] LHC *CERN-2004-003-V-1, LHC Design Report*, et al 2008 JINST 3 S08003
[2] ATLAS Collaboration, *The ATLAS Experiment at the CERN Large Hadron Collider*, JINST 3 (2008) S08003.
[3] CERN-LHCC-2012-022 ; *Letter of Intent for the Phase-II Upgrade of the ATLAS Experiment* LHCC-I-023
[4] CERN-LHCC-2011-012 ; *Letter of Intent for the Phase-I Upgrade of the ATLAS Experiment*LHCC-I-020
[5] ATLAS Collaboration, *ATLAS Muon Spectrometer Technical Design Report*,CERN-LHCC-97-022 (1997).
[6] ATLAS Collaboration, *New Small Wheel Technical Design Report*, CERN-LHCC-2013-006.
[7] Benhammou, Y. *"Precision Tracking with Small-Strip Thin Gap Chamber (STGC): from Test Beam to ATLAS NSW Upgrade."* Proceedings of The European Physical Society Conference on High Energy Physics — PoS(EPS-HEP 2013), 2014, doi:10.22323/1.180.0486.
[8] I. Giomataris et al., *MICROMEGAS: a high-granularity position-sensitive gaseous detector for high particle-flux environments, Nucl. Instr. Meth.* A 376 (1996) 29-35.
[9] T. Alexopoulos, *A spark-resistant bulk-micromegas chamber for high-rate applications*, et al. Nucl. Instrum. Methods A, 640 (2011), pp. 110-118
[10] Y. Kataoka,S. Leontsinis,c and K. Ntekas, *Performance studies of a micromegas chamber in the ATLAS environment*, 2014 JINST 9 C03016
[11] I. Manthos et al., The Micromegas Project for the ATLAS New Small Wheel, AIP Conf. Proc. 207
[12] Preliminary MMFE-8 Specification, ver 0.5, February 2015
[13] P. Gkountoumis, *"Level-1 data driver card of the ATLAS new small wheel upgrade compatible with the phase II 1 MHz readout scheme," 2016 5th International Conference on Modern Circuits and Systems Technologies (MOCAST), 2016*,pp. 1-4, doi: 10.1109/MOCAST.2016.7495115.
[14] T. Alexopoulos, D. Antrim, C. Bakalis, G. De Geronimo, P. Gkountoumis, G. Iakovidis, P. Moschovakos, V. Polychronakos, A. Taffard, *The VMM readout system, Nuclear Instruments and Methods in Physics Research Section A: Accelerators, Spectrometers, Detectors and Associated Equipment,* ISSN 0168-9002
[15] R. M. Coliban, S. Popa, T. Tulbure, D. Nicula, M. Ivanovici, S. Martoiu, L. Levinson and J. Vermeulen," *The Readout Controller for the ATLAS New Small Wheel*", JINST 11 (2016) no.02, C02069. doi:10.1088/17480221/11/02/C02069
[16] A. Caratelli, S. Bonacini, K. Kloukinas, A. Marchioro, P. Moreira, R. De Oliveira and C. Paillard, *"The GBTSCA, a radiation tolerant ASIC for detector control and monitoring applications in HEP experiments,"* JINST 10 (2015) no.03, C03034. doi:10.1088/17480221/10/03/C03034
[17] P. Moreira, J. Christiansen, K. Wyllie, GBTx manual, ver. 0.15, October 2016
[18] J Anderson, A Borga, H Boterenbrood, H Chen, K Chen, G Drake, D Francis, B Gorini, F Lanni, G Lehmann Miotto, L Levinson, J Narevicius, C Plessl, A Roich, S Ryu, F Schreuder, J Schumacher, W Vandelli, J Vermeulen, J Zhang, *FELIX: a High-Throughput Network Approach for Interfacing to Front End Electronics for ATLAS Upgrades*, J. Phys.: Conf. Ser. 664 082050 May 2015.
[19] A. Zibell, *High-Rate Irradiation of 15mm Muon Drift Tubes and Development of an ATLAS Compatible Readout Driver for Micromegas Detectors*, Dissertation, LMU Munich (2014)
[20] E. Karentzos *The HV DCS System for the New Small Wheel Upgrade of the ATLAS Experiment* 18th Int. Conf. on Acc. and Large Exp. Physics Control Systems
[21] CERN Gamma Irradiation Facility, *The Gamma Irradiation Facility GIF++ during and beyond Run 3*
[22] T. Alexopoulos et al., *Performance studies of resistive-strip bulk micromegas detectors in view of the ATLAS New Small Wheel upgrade*, Nuclear Inst. and Methods in Physics Research, A 937 (2019) 125–140]
[23] M. Bianco, *Micromegas detectors for the muon spectrometer upgrade of the ATLAS experiment Micromegas detectors for the muon spectrometer upgrade of the ATLAS experiment*, Nuclear Instruments and Methods in Physics Research Section A: Accelerators, Spectrometers, Detectors and Associated Equipment Volume 824, 496-500

Segment Linking: A Highly Parallelizable Track Reconstruction Algorithm for HL-LHC

P Chang[1], P Elmer[2], Y Gu[1], V Krutelyov[1], G Niendorf[3], M Reid[3], B V Sathia Narayanan[1], M Tadel[1], E Vourliotis[1a], B Wang[2], P Wittich[3], A Yagil[1]

[1] University of California San Diego, CA, US
[2] Princeton University, NJ, US
[3] Cornell University, NY, US

E-mail: [a] emmanouil.vourliotis@cern.ch

Abstract. The High Luminosity upgrade of the Large Hadron Collider (HL-LHC) will produce particle collisions with up to 200 simultaneous proton-proton interactions. These unprecedented conditions will create a combinatorial complexity for charged-particle track reconstruction that demands a computational cost that is expected to surpass the projected computing budget using conventional CPUs. Motivated by this and taking into account the prevalence of heterogeneous computing in cutting-edge High Performance Computing centers, we propose an efficient, fast and highly parallelizable bottom-up approach to track reconstruction for the HL-LHC, along with an associated implementation on GPUs, in the context of the Phase 2 CMS outer tracker. Our algorithm, called Segment Linking (or Line Segment Tracking), takes advantage of localized track stub creation, combining individual stubs to progressively form higher level objects that are subject to kinematical and geometrical requirements compatible with genuine physics tracks. The local nature of the algorithm makes it ideal for parallelization under the Single Instruction, Multiple Data paradigm, as hundreds of objects can be built simultaneously. The computing and physics performance of the algorithm has been tested on an NVIDIA Tesla V100 GPU, already yielding efficiency and timing measurements that are on par with the latest, multi-CPU versions of existing CMS tracking algorithms.

1. Introduction and Motivation

The Large Hadron Collider (LHC) of CERN is currently the largest accelerator in the world. In order to take advantage of the extensive infrastructure for its operation for as long as possible, a major upgrade to the LHC is planned for 2027 (Phase 2). This upgrade, called the High Luminosity LHC (HL-LHC), aims at collecting more than $3000\,\text{fb}^{-1}$ of integrated luminosity of proton-proton (p-p) collision data, increasing the corresponding nominal value of the LHC by a factor of ten [1]. To achieve such high values of integrated luminosity, the instantaneous luminosity, i.e. rate of collisions, is pushed to its limits. This leads to an increased pileup (PU), i.e. number of simultaneous p-p interactions per bunch crossing. The average PU during the full LHC operation (including Run 3) is expected to stay under 70, while the same number for the HL-LHC will rise to around 200.

The much harsher PU conditions anticipated in the HL-LHC lead to an escalation of the complexity of event reconstruction. Especially for the reconstruction of charged-particle tracks, which relies on the combination of multiple hits and is an inherently combinatorial problem, several times more PU results in unprecedented tracker occupancies and computational complexity. As shown in reference [2], this translates to an exponential growth of time needed for the track reconstruction. To mitigate this effect, increased computational resources would be required, leading to highly elevated costs of operation.

Despite the projected increase in the processing power of central processing units (CPUs), which dominated the computational resources of Run 1 and 2 of the LHC, the computational complexity of event reconstruction at the HL-LHC is expected to demand a computational cost that exceeds the computing budget using conventional single-thread CPUs programming. As a result, alternative approaches need to be explored. A solution to this problem is the efficient parallelization of existing reconstruction algorithms, where this is feasible, or the creation of new algorithms that can benefit from parallelism. The concept of parallelization can be combined with the modern paradigm of heterogeneous computing, which aims at utilizing a large variety of processing units, such as graphics processing units (GPUs) or field-programmable gate arrays (FPGAs).

This paradigm shift is already implemented in Run 3 related applications in experiments of the LHC. Taking the example of the Compact Muon Solenoid (CMS) experiment [3], the pixel track and vertex reconstruction (Patatrack [4]), the outer tracker strip local reconstruction, and the electromagnetic calorimeter and hadronic calorimeter reconstruction algorithms have been ported to GPUs. The combination of CPU and GPU usage leads to $\sim 25\%$ timing reductions for the total Run 3 CMS High Level Trigger (HLT) reconstruction. The HL-LHC cost projections show that offloading 50% (80%) of the computational work to GPUs decreases the HLT processing farm cost by 35% (75%) [5].

Moving in this direction, this work presents a highly parallelizable algorithm for track reconstruction at the HL-LHC, called *Segment Linking* or *Line Segment Tracking (LST)*. The algorithm has been developed to benefit from heterogeneous computing architectures, more specifically from GPU performance, and a relevant implementation of it on an NVIDIA Tesla V100 GPU is described.

2. The Segment Linking Algorithm

The track reconstruction at the LHC is usually based on Kalman filter methods [6], which are inherently sequential in nature. This has repercussions in terms of computational time for their application. For example, for the CMS experiment, track reconstruction used to require almost 60% of the time needed for the whole event reconstruction procedure at 50 PU. There has been a recent effort towards the parallelization of Kalman filter approaches (*mkFit* project [7]) that improves the timing performance with comparable physics results. Nevertheless, there is a clear need for track reconstruction algorithms with architectures specifically designed for parallelization.

LST is an algorithm developed with the concept of parallelization built into it. The algorithm relies on local hits in the tracker to start producing short tracks. Given the local nature of these short tracks, the algorithm utilizes the Single Instruction, Multiple Data (SIMD) paradigm, where all of the short tracks can be created concurrently with the same set of instructions, since they are independent from one another. Progressively, the short tracks are linked together to form longer tracks, leading to collections of objects with different characteristics. Finally, objects from these collections are combined to create a Track Candidate collection with a high efficiency and a low fake rate. LST is inspired by the XFT algorithm in the Collider Detector at Fermilab (CDF) at the Tevatron [8] and its prototype was presented at the International Conference of High Energy Physics (ICHEP) in 2016 [9].

Before delving into the details of the track building logic of the algorithm, it is worth noting a real example for its application. The CMS Phase 2 Outer Tracker, shown in figure 1 (blue and red lines) is a suitable detector for the LST algorithm. This comes from the fact that each layer is composed by multiple "p_{T} modules" (lines in figure 1), each with two closely-spaced sensors which can provide a very localized combination of hits, called *stub*, as shown in figure 2. In the LST algorithm presented here, a stub is accepted only if the hits are within a selection window (marked in green in figure 2) which corresponds to a p_{T} threshold of 0.8 GeV for the traversing particle. The usage of stubs instead of individual hits can significantly help in reducing the combinatorics in the initial steps of the LST algorithm up to 7.5 times. As an example, the number of hits in the first layer of the outer track in one $\mathrm{t\bar{t}}$ event at 200 PU event is approximately 36000, while this number goes down to 6000 for stubs in the same layer.

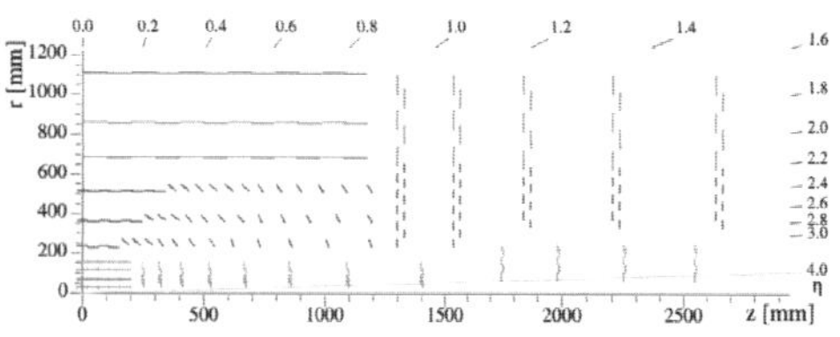

Figure 1. $r - z$ perspective of one quarter of the Phase-2 CMS tracker. The green and orange lines represent the inner tracker modules, while the blue and red lines represent the outer tracker modules [10].

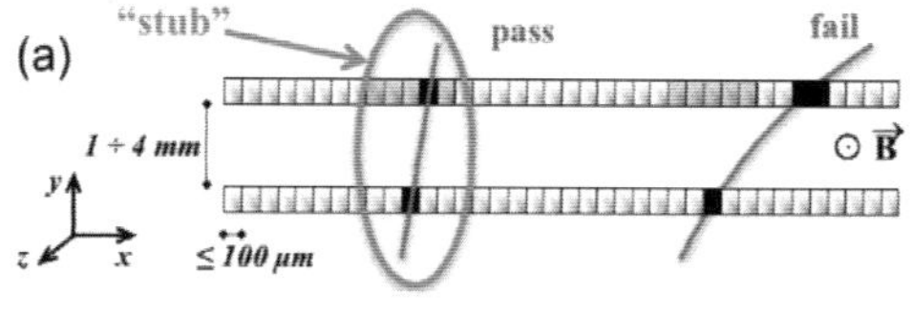

Figure 2. Sketch of a tracker p_{T} module. The series of gray squares represent the closely-spaced sensors of a module, while the green squares indicate the channels compatible with a particle which has passed through reference channel (black square). The lower the particle p_{T}, the lower the probability to produce an accepted stub [10].

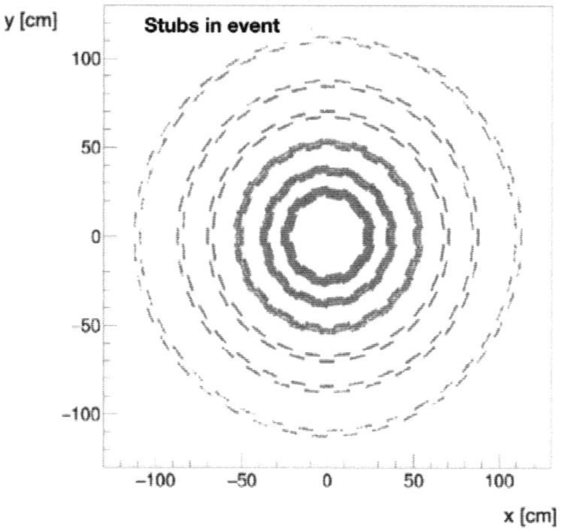

Figure 3. Occupancy distribution for stubs in a single $\mathrm{t\bar{t}}$ event at 200 PU in the barrel part of the Phase-2 CMS tracker as a function of the x and y coordinates of the CMS detector.

2.1. The Logic of the Algorithm

The track building procedure starts with the creation of stubs in each of the p_{T} module of the outer tracker. The occupancy distribution of stubs in the $x - y$ two-dimensional plane of the CMS coordinate system is shown in figure 3 for a typical $\mathrm{t\bar{t}}$ event at 200 PU. The next step is the linking of stubs in neighboring layers (figure 6) together to form a *segment* (figure 7). With around 6000 stubs in a single layer, the combinatorics for all the stub connections is immense. To decrease the number of combinations, physics requirements are taken into account. More specifically, helices of particles originating from the center of the detector, with both positive and negative sign hypotheses and with p_{T} at the previously mentioned threshold of 0.8 GeV, are considered. Tracks from simulation are considered as well, in order to include cases that may have been missed by the idealistic physics scenario of helices. This leads to restrictions in the number of modules that can be linked to a single (reference) module, as it can be seen in figure 4.

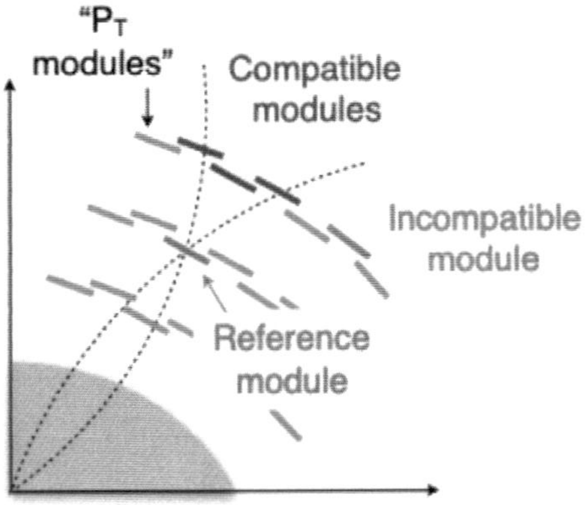

Figure 4. Demonstration of the logic behind the construction of a map of modules compatible for linking algorithm objects. The dashed line represents helices of positively and negatively particles with $p_{\mathrm{T}} = 0.8\,\mathrm{GeV}$.

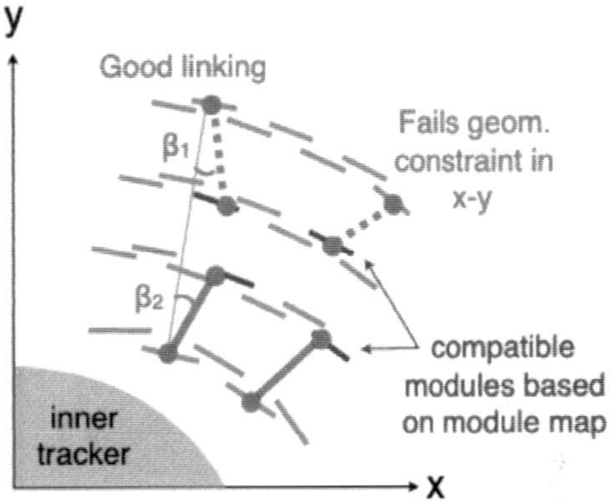

Figure 5. Two examples showcasing how geometrical requirements, in combination with the list of compatible modules, lead to accepting or rejecting the linking of algorithm objects.

The creation of segments is followed by the creation of *triplets (T3)* and *quintuplets (T5)*. Triplets are formed from the linking of two segments that share a common stub (figure 8), while quintuplets involves the linking of two triplets with a common stub (figure 9). As objects get longer, more handles can be utilized for determining whether their linking is consistent with a track hypothesis. As an example, figure 5 shows that the relative angle between segments can be used to reject their combination if that leads to an unphysical track pattern, such as the right set of segments in the figure.

In addition to the outer-tracker-only objects, exploiting information from the inner tracker can assist in creating collections with higher purity of physical tracks. Triplets and quintuplets are linked to seeds from the pixel detector, called *pixel line segments (pLS)*, hence producing even longer objects, *pixel triplets (pT3)* (figure 10) and *pixel quintuplets (pT5)* (figure 11), respectively.

After all of the objects have been created, a "cross-cleaning" procedure is applied. This ensures that duplicates from different object collections are removed from the final output *track candidate (TC)* collection and is implemented as follows: First, any pixel triplets that are close to any pixel quintuplet in the $\eta - \phi$ plane are marked as duplicates. The same procedure then takes place for quintuplets versus both pixel quintuplets and cleaned pixel triplets. Finally, remaining pixel line segments, unused in the previous linking steps, are cross-cleaned versus all the other objects. In the above cross-cleaning steps, it is also required that objects do not share any hits. As the last step of the algorithm, the track candidate collection is created by adding to it all of the following, non-duplicate objects:

- pixel quintuplets,

- pixel triplets,
- quintuplets, and
- unlinked pixel line segments.

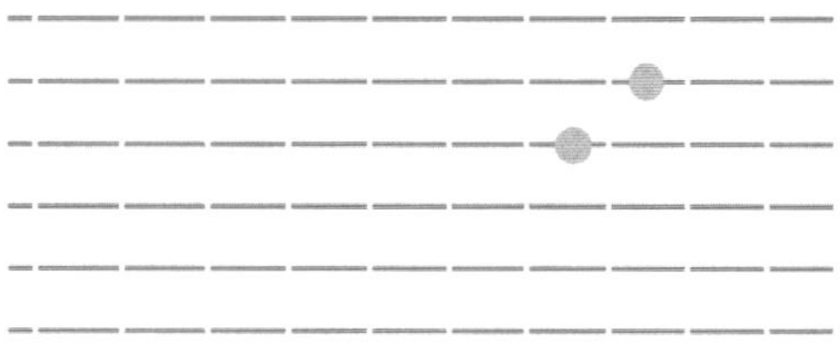

Figure 6. Stubs.

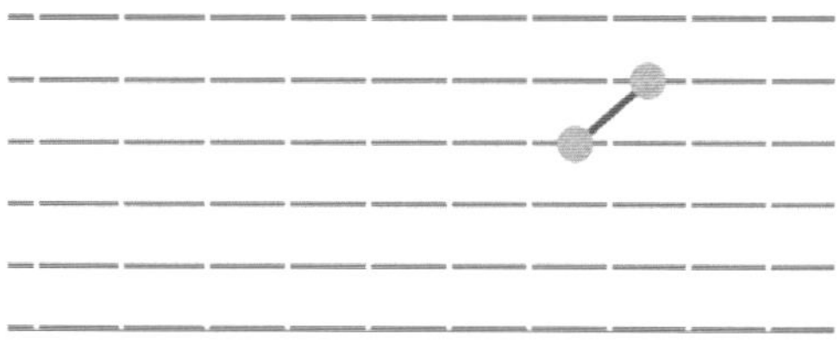

Figure 7. Segment.

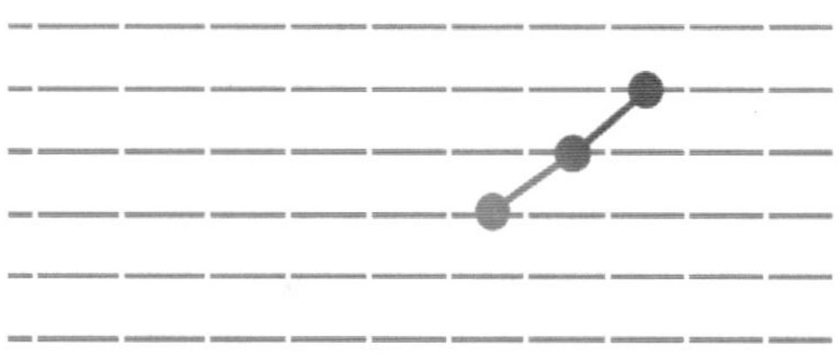

Figure 8. Triplet.

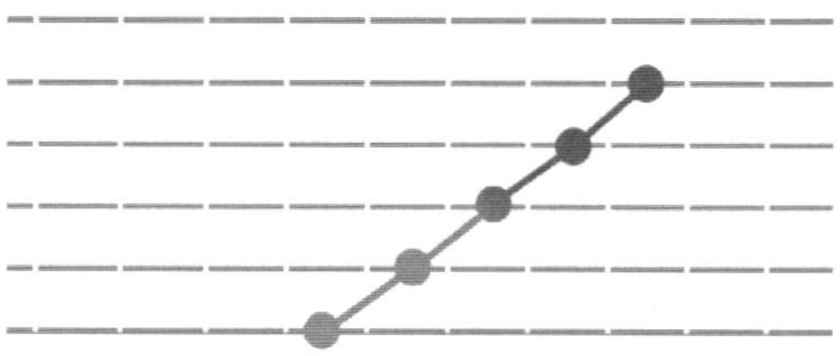

Figure 9. Quintuplet.

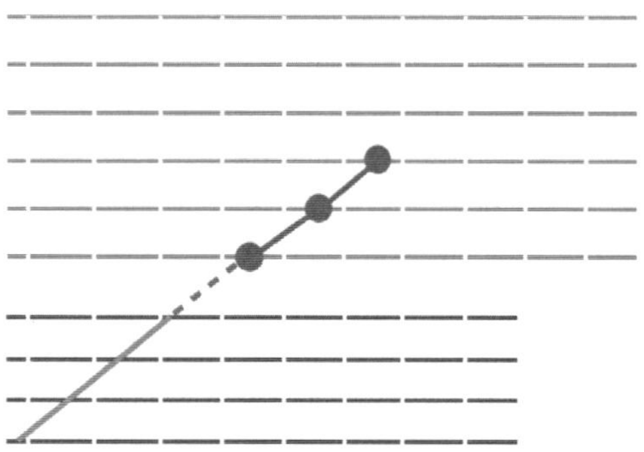

Figure 10. Pixel triplet.

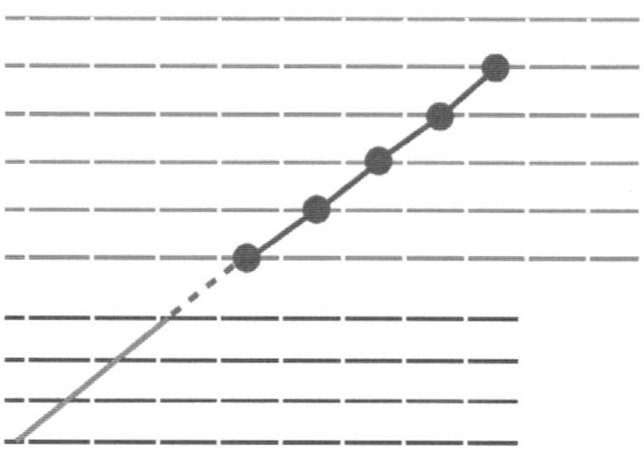

Figure 11. Pixel quintuplet.

The different objects created in the LST algorithm. The cyan lines represent the outer tracker layers, while the dark red lines represent the inner tracker layers.

2.2. Physics Performance

The selection of the objects that comprise the final track candidate collection is based on maximizing the track finding efficiency with the lowest possible fake rate. Each different object category aims at recovering efficiency of tracks with different characteristics. This is showcased in figures 12-14, which show the breakdown of the track candidate efficiency (TC, in black) in different objects as a function of different variables.

Starting from figure 12, which presents the efficiency as a function of p_{T}, it is evident that the pixel quintuplets (pT5, in red), which are the longest objects, covering almost the whole range of the detector layers, are the driver of the performance. The same figure indicates that the pixel triplets (pT3, in green) and unlinked pixel line segments (pLS, in magenta) recover some efficiency at lower p_T. This is because low p_T tracks have a lower probability of leaving hits in a lot of tracker layers. The total efficiency for track candidates plateaus at its maximum value of 90% already at 5 GeV. The real benefit of including the unlinked pixel line segments in the track candidate collection reveals itself in the efficiency plot as a function of η, where they noticeably are the drivers of the performance at the high $|\eta|$ range. Finally, figure 14 shows that the inclusion of quintuplets (T5, in blue) is the unique handle of reconstructing displaced tracks.

In terms of fake rate, figures 15 and 16 present it as a function of p_T and η respectively. Above the lower p_T threshold of 0.8 GeV, the fake rate remains less than 10% for the bulk of the events, up till ~5 GeV. For higher p_T values, some requirements on the objects are loosened to improve efficiency, which leads to an increase in the fake rate. The fake rate tends to be higher in the central region, where less accurate p_T modules are employed with respect to the region of $|\eta| > 2.2$. Due to the extensive cross-cleaning, the duplicate rate is less than 5%, flat as a function of p_T, with the main contribution coming from unused pixel segments at high $|\eta|$.

As a reference for the performance of the LST algorithm, the current CMS algorithms, adapted for HL-LHC, can be used. Looking at the "Patatrack Trimmed" configuration of reference [11], the efficiency of the LST algorithm is comparable with that, consistently around 90% over the whole $|\eta|$ range. At the same time, the LST fake rate is higher, reaching ~15% versus 4% in the barrel and ~20% versus 14% in the endcap. It is worth mentioning that the comparison can only be very rough, since important components of track reconstruction, such as the track final fitting, have not been implemented yet for the LST algorithm.

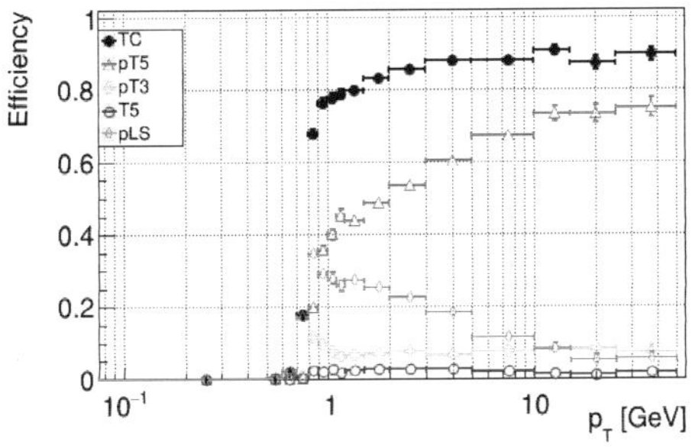

Figure 12. Efficiency for reconstructed tracks in a $t\bar{t}$ sample at 200 PU, as a function of the generator track p_T. The color code is explained in the text.

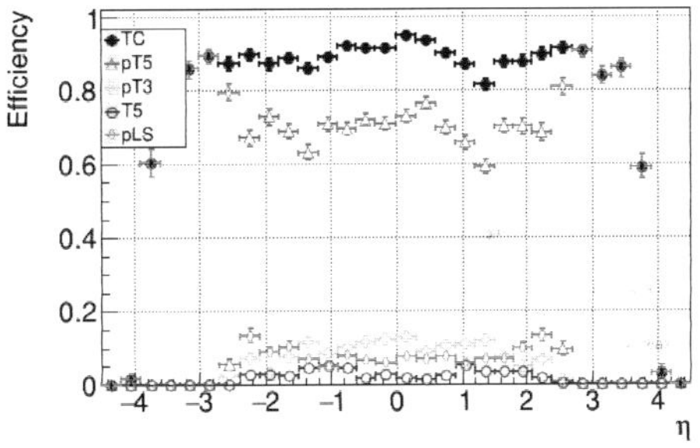

Figure 13. Efficiency for reconstructed tracks in a $t\bar{t}$ sample at 200 PU, as a function of the generator track η. The color code is explained in the text.

3. GPU Implementation

GPUs have both advantages and disadvantages when compared to CPUs, making each one of these processing units suitable for different applications. GPUs host a much larger number of cores, $\mathcal{O}(10^3)$, than CPUs do, which have at most a few tens of cores. Due to this, GPUs can achieve greater numbers of floating point operations per second, hence being able to maintain high throughput. On the other hand, GPUs have limited memory resources, which can lead to high latency, when host-to-device data transfers are needed, while CPUs have lower memory latency, owing to more extensive use of caches. Given these characteristics, CPUs are more suitable for serial processing, while GPUs excel in parallel computations.

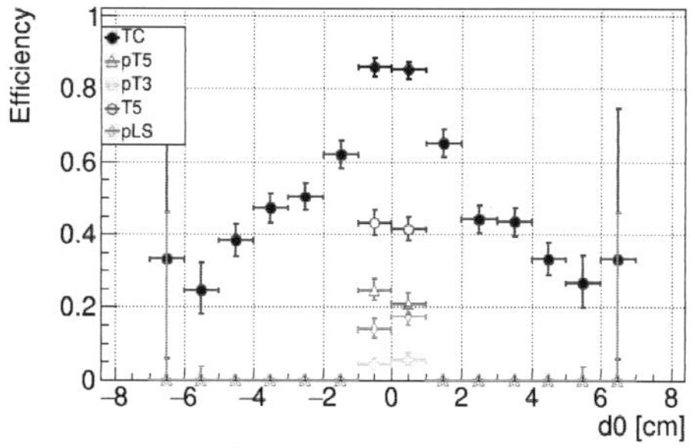

Figure 14. Efficiency for reconstructed tracks in a muon gun sample at 0 PU with the muon production vertex uniformly distributed in a cube with edge of 5 cm, as a function of the generator track d_0, i.e. its distance from the center of the detector. The color code is explained in the text.

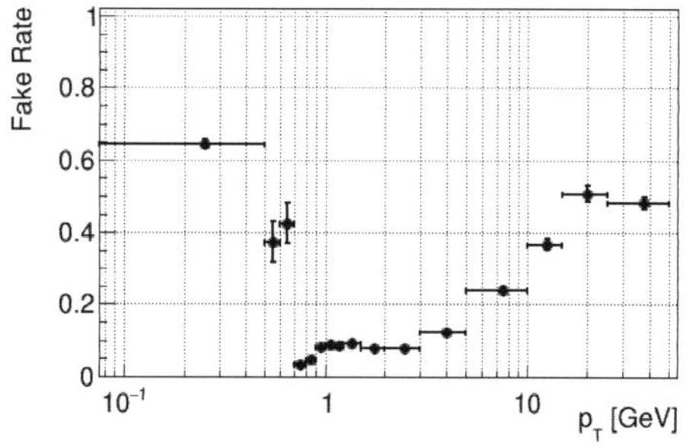

Figure 15. Fake rate for reconstructed tracks in a $t\bar{t}$ sample at 200 PU, as a function of the reconstructed track p_{T}.

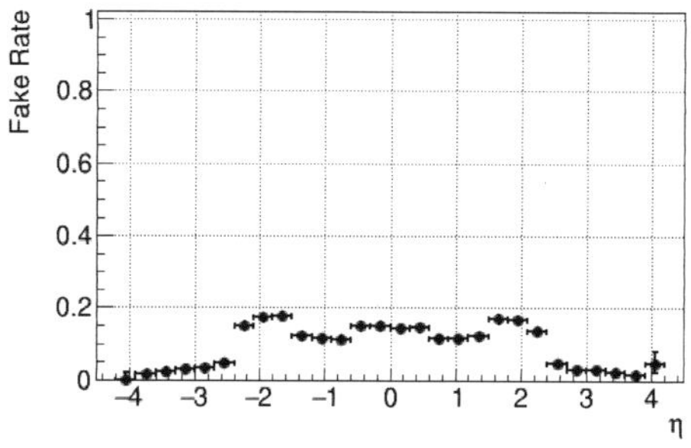

Figure 16. Fake rate for reconstructed tracks in a $t\bar{t}$ sample at 200 PU, as a function of the reconstructed track η.

From the considerations above, it is clear that GPU programming requires both a parallelizable algorithm and an appropriate implementation to exploit the full capabilities of GPUs. As already mentioned, LST is an algorithm that is fit for parallelization according to the SIMD principle and our implementation, described below, aims to take advantage of the multitude of GPU cores while efficiently handling memory assignments and transfers. The LST algorithm implementation on GPU has been developed with CUDA, a programming framework for general-purpose computing on GPUs, developed by NVIDIA.

One cornerstone of the LST algorithm implementation is the usage of Structure of Arrays format, commonly referred to as "SoA", with explicit memory management. Instead of using the common practice of object-oriented programming to create classes that hold the properties of a single object, separate arrays are created for separate object properties. These arrays hold each relevant quantity in continuous memory addresses, hence allowing for successive memory reads of the same quantity to be contiguous (memory coalescing). In this way, the time needed for accessing memory is minimized. Additionally, a custom cache memory allocation is utilized for small and commonly used objects. The cache memory is of very limited size but it allows for extremely fast memory transfers, helping further to decrease the total execution time of the algorithm. Finally, extensive studies on the final occupancy of all algorithm objects have led to truncations of their output numbers at the level of 99.9% or 99.99%, resulting in a 5-fold reduction of the total memory footprint for the algorithm, while sacrificing only 1–2% of efficiency.

In terms of organization of its algorithmic steps, the LST code is split into different kernels for the creation of each object. A kernel is a set of operations that can run in parallel on independent GPU threads. Due to the fact that inputs for each kernel are highly localized, a large number of independent object reconstruction threads can run in the same kernel concurrently. The order in which kernels are launched roughly follows the description of the algorithm in subsection 2.1 and is shown in figure 17.

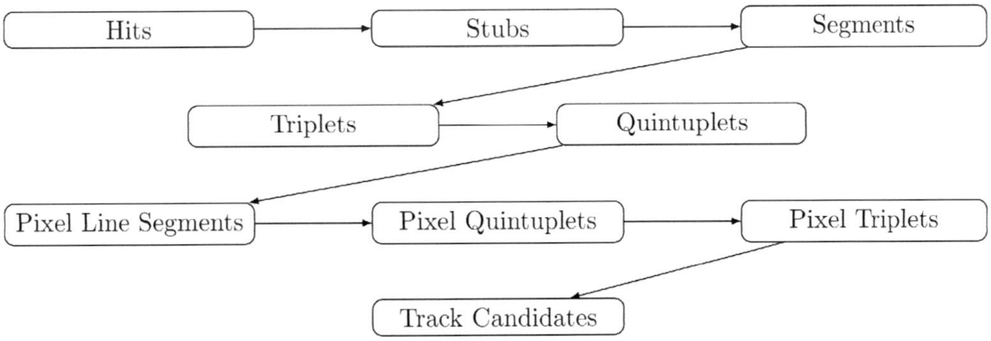

Figure 17. Flowchart of the order of kernel execution.

There is another layer of parallelization that can be exploited: event-level parallelization. Instead of running only one event in one GPU stream, multiple streams can be employed, each one of which runs a different event. This technique, called "multi-streaming", allows for better utilization of the GPU resources and provides more opportunities for the kernel scheduler to insert more work. Moreover, it can hide the high GPU latency by assigning work to processing elements which already have data available, while others wait for data to be transferred. This can be seen in figure 18, where light blue indicates the total work done by the GPU and dark blue shows the work done by kernels in different streams, both as a function of time. In the 1-stream case (upper schematic), the total GPU work is interrupted while the single kernel running waits for new data. On the other hand, in the 8-stream case (lower schematic), even though kernels in separate stream have similar waiting times, the total work done by the GPU is continuous in time, as data loading is hidden by work in other streams.

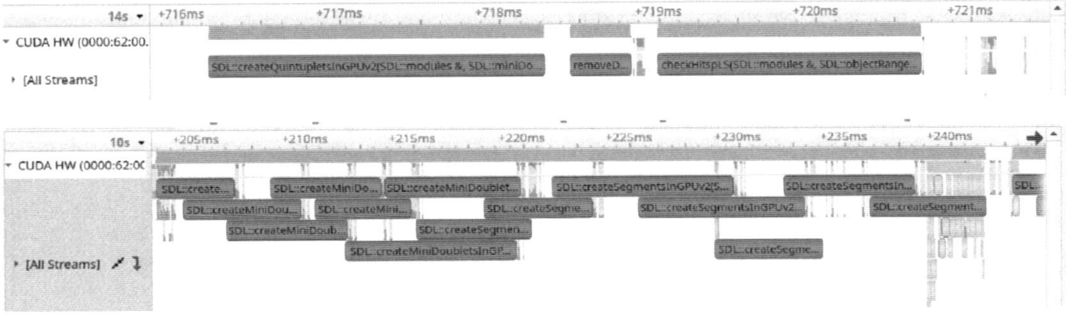

Figure 18. Comparison of algorithm execution on 1-stream (upper) versus 8-streams (lower), using the NVIDIA Visual Profiler. Multi-streaming allows for running parts of multiple kernels concurrently.

The benefits of multi-streaming are evident when measuring the time needed for the execution of the LST algorithm. This timing measurement is performed in a $t\bar{t}$ sample at 200 PU and excludes the initial data transfer host-to-device for the hits, as well as any final fitting procedure on the output tracks, which has not been implemented in the current version of the algorithm. The average time needed for the execution of the algorithm on a single event is 32 ms for the 1-stream case, while it decreases to 26 ms for the 8-stream case, resulting in $\sim$20% improvement with respect to the 1-stream case. These timing measurements are in the same range as the latest CMS tracking timing measurements on CPUs [11]. It is worth noting that the LST algorithm is on par also price-wise with multi-CPU efforts, since two 32-core Intel Skylake Gold Xeon processors, commonly used in such efforts, cost approximately the same as one NVIDIA Tesla V100 GPU that has been used in our case.

4. Summary and Outlook

In this work, Segment Linking (or Line Segment Tracking), a highly parallelizable track finding algorithm, aimed at alleviating the challenges that the HL-LHC will pose for tracking, has been presented. The algorithm has been successfully implemented on an NVIDIA Tesla V100 GPU and has been shown to have both physically and computationally comparable performance with cutting edge tracking efforts on CPUs.

More improvements of the algorithm are planned for the future. On the physics side, a re-evaluation of the selection criteria for objects is underway to improve the efficiency and lower the fake rate, especially for displaced tracks. Integration with other, similar GPU developments, such as Patatrack, is also foreseen. Computationally, mathematical optimizations for the calculation of physics parameters, as well as the extension of the usage of alternative data types, e.g. half-precision floats, are being explored. Plans to refine memory coalescing will potentially help reducing timing even further. Ultimately, the grand design for the algorithm involves its integration to the central CMS software for HLT and offline usage during the HL-LHC operation.

5. Acknowledgements

This work was supported by the U.S. National Science Foundation under Cooperative Agreements OAC-1836650 and PHY-2121686 and grant NSF-PHY-1912813.

References

[1] Aberle O et al 2020 High-Luminosity Large Hadron Collider (HL-LHC): Technical design report CERN-2020-010 URL `https://cds.cern.ch/record/2749422`
[2] Cerati G B 2014 Tracking and vertexing algorithms at high pileup CMS-CR-2014-345 URL `https://cds.cern.ch/record/1966040`
[3] The CMS Collaboration 2008 *Journal of Instrumentation* **3** S08004
[4] Bocci A, Innocente V, Kortelainen M, Pantaleo F and Rovere M 2020 *Frontiers in Big Data* **3**
[5] The CMS Collaboration 2021 The Phase-2 Upgrade of the CMS Data Acquisition and High Level Trigger CERN-LHCC-2021-007, CMS-TDR-022 URL `https://cds.cern.ch/record/2759072`
[6] Frühwirth R 1987 *Nuclear Instruments and Methods in Physics Research Section A: Accelerators, Spectrometers, Detectors and Associated Equipment* **262** 444
[7] Cerati G, Elmer P, Gravelle B, Kortelainen M, Krutelyov V, Lantz S, Masciovecchio M, McDermott K, Norris B, Reid M, Hall A R, Riley D, Tadel M, Wittich P, Würthwein F, Yagil A 2019 Speeding up Particle Track Reconstruction in the CMS Detector using a Vectorized and Parallelized Kalman Filter Algorithm URL `https://arxiv.org/abs/1906.11744`
[8] Ciobanu C, Gerstenslager J, Hoftiezer J, Hughes R, Johnson M, Koehn P, Neu C, Sanchez C, Winer B L, Freeman J, Holm S, Lewis J, Shaw T, Wesson T, Bloom K, Gerdes D, Dawson, J W, Haberichter W N 1999 *IEEE Transactions on Nuclear Science* **46** 933
[9] Krutelyov V, Cerati G, Tadel M, Wuerthwein F, Yagil A 2017 *Journal of Physics: Conference Series* **898** 042023
[10] The CMS Collaboration 2017 The Phase-2 Upgrade of the CMS Tracker CERN-LHCC-2017-009, CMS-TDR-014 URL `https://cds.cern.ch/record/2272264`
[11] The CMS Collaboration 2021 Performance of Phase-2 HLT Reconstruction and GPU offloading benchmarks CMS-DP-2021-013 URL `https://cds.cern.ch/record/2792313`

Journal of Physics: Conference Series **2375** (2022) 012006 doi:10.1088/1742-6596/2375/1/012006

The Innovative FEL design by the CompactLight Collaboration

N Gazis[1] A Bignami[1] E Trachanas[1] T Alexopoulos[2] I Telali[2] T Apostolopoulos[4] K Pramatari[4] A Karagiannaki[4] D Kotsopoulos[4] E Gazis[2,3] and D V Bandekas[5]

[1] European Spallation Source ESS-ERIC, Lund Sweden
[2] National Technical University of Athens, Athens, Greece
[3] Institute of Accelerating Systems and Applications, Athens, Greece
[4] Athens University of Economics and Business, Athens, Greece
[5] International Hellenic University, Physics Department, Kavala, Greece

Email: nick.gazis@ess.eu

Abstract. The innovative FEL design by the CompactLight Collaboration is briefly presented. The major contribution of our team, mainly from Greece cooperating with the team of ESS-ERIC, is emphasised to the material selection of the proper photocathode illuminated with the relevant laser, the magnet and solenoid design of the injector for optimum beam conditions, the 3D CAD model design of the baseline FEL facility including the machine, the tunnel, the beam dump and the experimental hall. Additionally, the exploitation assets of the project with the transfer technology are presented.

1. Introduction

The successful use of many Synchrotron Radiation facilities for the recent previous decades to applications to the various scientific and engineering disciplines has, after that, led to the explosion of the Free Electron Laser facilities driven by linacs [1]. The FEL facilities provide sub-picosecond pulse-lengths and wavelengths down to the hard X-ray range, innovative performance of the peak brightness, exceeding by many orders of magnitude that of existing generation synchrotrons and enabling important complementary research opportunities.

The CompactLight Collaboration an H2020 design study project, funded by the European Commission, aims to facilitate the widespread development of X-ray FEL facilities across Europe and beyond, by making them more affordable to construct and operate, through an optimum combination of emerging and innovative accelerator technologies.

The CompactLight FEL has provided specifications, driven by the demands of its potential users and the X-rays characteristics required by the experiments. The Conceptual Design Report (CDR) [2], delivered by the CompactLight Collaboration has been based on the latest concepts for bright electron photoinjectors, high-gradient X-band structures at 12 GHz, and innovative short-period undulators.

The exploitable assets of each work package of the CompactLight project have been collected and studied provided the current and future possibilities by the industry or the academia institutions to evaluate the benefits of their investment to the innovative FEL building facility or its components. The innovative accelerating technology plus additional components of the CLIC Collaboration [3] have been taken as knowledge transfer to the CompactLight Collaboration, who has setup a coherent FEL design.

2. The CompactLight FEL design

The CompactLight FEL design has been achieved by the adoption of innovative technical solutions, in respect to the existing facilities under operation, for the same working wavelengths, with a lower electron beam energy. A significantly more compact footprint, in case of construction, the total length of the facility is just over 480 m, which is, instead of the 730m total length of the SwissFEL [4]. In brief the list of the key elements considered for the design study are presented.
The high FEL stability in pulse energy and pulse duration with the FEL synchronization is better than 10 fs. The photon pulse duration is less than 50 fs with a repetition rate from 1 Hz up to 1 kHz. The focused spot size is small with variable polarization, linear and elliptical. It is provided a tunability up to higher photon energies with two-bunch operation and two-colour pulse generation.
The CompactLight FEL has been designed as a hard X-ray facility, spanning the wavelength range from 0.8 Å up to 5nm (16 keV to 0.25 keV) with two separate FEL beamlines:

i) a soft X-ray (SXR) FEL providing photons from 5.0nm to 0.6nm (0.25 keV to 2 keV) operating up to 1 kHz repetition rate (high rep rate);
ii) a hard X-ray (HXR) FEL providing photons from 6.0Å to 0.8 Å (2 keV to 16 keV) with maximum 100 Hz repetition rate (low rep rate).

The dual-bunch photoinjector and the two-beam deflectors adopted for the linac in the design, having a huge flexibility for the facility operation, with different combinations of SXR and HXR operating modes, at high and low repetition rates, as requested by the users.
The CompactLight FEL, also, includes a soft X-ray FEL and an extremely compact and relatively inexpensive photon source based on Inverse Compton Scattering (ICS), both using the CompactLight technology. This soft X-ray FEL can be considered a quite affordable solution, in term of cost and complexity, in case of limited funding capabilities. The ICS source, with its wide range of applications, is very attractive and can be easily installed and operated in university campuses, small laboratories, and hospitals.

3. The FEL components

The major components of the CompactLight FEL are described with the main technical specifications

3.1. The electron source and injector

The electron source has a design of an innovative C-band photoinjector, with an operating gradient up to 180 MV/m, that can deliver two 75 pC e-bunches per RF pulse, with less than 0.2 mm mrad normalized emittance.
The main criteria for the selection of the cathode have been the high quantum efficiency, high robustness and fast response time in addition to low intrinsic emittance, low surface roughness and high lifetime. Moreover, the photocathode is also subject to surface modification and contamination due to the photoemission. The metal photocathodes require UV light, generated from 3rd or 4th harmonic conversion from an IR fundamental wavelength. The injector is using a gun and a photoinjector laser, to fulfill the FEL operating modes, able to operate with a two e-pulses, for each RF pulse, spaced by 3 or 5 RF cycles of the C-band frequency a Copper Cathode driven by a 1 kHz Ti:Sa Laser is the choice for the electron source. The remaining part of the booster linac, up to the first bunch compression, at 300MeV, consists of a full C-band linac.
The development of very high frequency linearizers is of broad importance for accelerators which require short beam bunches, including high frequency RF and plasma accelerators. For the CompactLight FEL longitudinal phase space linearization, a Ka-band system has been designed

operating at 36 GHz. The Ka-band system is based on a 300mm travelling-wave structure, powered with 3MW RF and operating with an integrated gradient of 12.7MV.
The injector system extends from the electron source to the Bunch Compressor 1 (BC-1) exit, corresponding to an upper energy of 300MeV. At about 300MeV the electron beam experiences a transition from the "space-charge" dominated regime to the "emittance dominated" regime and the emittance compensation process, performed within the injector itself, can be considered accomplished. The injector parameters are presented below in Table 1:

Table 1

Parameter	Unit	After VB and/or BC-1
Charge Q	pC	75
Beam energy	MeV	300
RMS Bunch Duration σ_t	fs	350
Peak Current	A	60
RMS Energy Spread	%	0.5
Projected RMS Norm. Emittance	mm	0.2
Repetition Rate	Hz	100–1000

Table 1: The basic injector parameters are presented, providing the excellent conditions of the beam coming out of the e-gun.

The RF gun is designed with mode launcher and solenoid. The solenoid around the circular waveguide has been designed using Poisson Superfish [5]. The simulated structure and the magnetic field profiles on axis are given in Figure 1. [6]

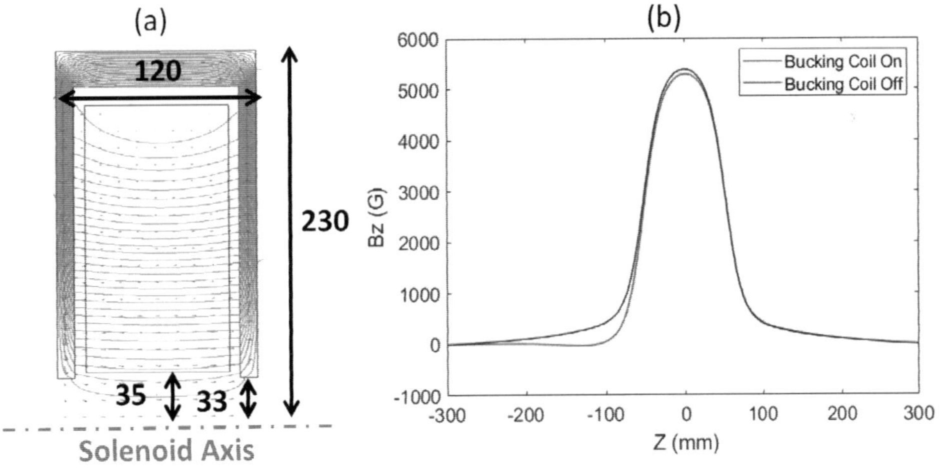

Figure 1.: (a) Solenoid simulated by the code Poisson and magnetic field profile on axis (b)

The design adopts a metallic photocathode to satisfy the CompactLight requirements. This choice is based on the results obtained worldwide in different laboratories, and on the ASTRA [7] simulation studies, comparing the transverse emittance and the normalized transverse emittance between the cathode materials Cs_2Te vs Cu. The S-/C-band of the e-gun has been studied, with a solenoid after the photocathode, at the Ferrario working point. The proper laser parameters have been studied for final selection of wavelength, pulse duration, pulse-length, pulse energy, intrinsic emittance and QE tuning, plus the repetition rate, the longitudinal-transverse and 3D pulse profile optimization. The GIOTTO [8] code was implemented and provided optimization results of the normalized transverse emittance [9]. Simulations have been done also to optimize transverse/longitudinal emittance, beam size, bunch length,

with 1.6, 2.5 and 5.6 cavity cells, with various cavity voltages and solenoid distances from the photocathode [10]. The most suitable candidate metal was found to be copper (Cu) [10]. The photocathode is centred on the flange and closes the RF gun half-cell.
The detailed results of our simulation studies for the two photocathode candidate materials are summarized in the Tables 2 and 3; taking into consideration the two cases of the e-guns 1.5 and 2.5 shells respectively.

Table 2

MATERIAL	Cs_2Te	Cu
Charge (pC)	75	75
Radius rms (mm)	0.17	0.17
Laser λ(nm)/E(eV)	262nm/4.73eV	262nm/4.73eV
Laser pulse energy(μJ)	35.48	0.221
Laser pulse duration (psec)	10	10
Laser rise time/fall time (psec)	0.5	0.5
Work function (eV)	3.5	4.65
Φ_{exc} (eV)	1.23	0.08
Electric field on cathode (MV/m)	100	100
$\Phi_{schottky}$ (eV)	0.379	0.379
Delay (psec)	0.4	0.0
Distribution	Isotropic	Isotropic
Normalised Emittance (mm mmrad/mm)	0.90	0.23
Normalised Emittance-schottky (mm mmrad/mm)	1.02	0.55
Emittance-schottky (mm mmrad/mm)	0.1734	0.0935

Table 3

e-GUNS	No 1	No 2
Number of shells	1.5	2.5
Maximum electric field (MV/m)	100	100
e-Gun phase (degrees)	215.3	237
Solenoid number of coils	1	1
Solenoid max. magnetic field (T)	0.2079	0.2079
Solenoid position (m)	0.3	0.3
Final position of tracking (m)	0.6	0.6

Table 2: The most characteristic specifications are presented for the two photocathode materials Cs_2Te and Cu.
Table 3: The conditions of the e-guns with 1.5 and 2.5 shells respectively are presented, under which there have been done the simulations studies.

An example of our transverse emittance simulation studies is presented in the figure 2 with the two e-guns and the two candidate photocathode materials.

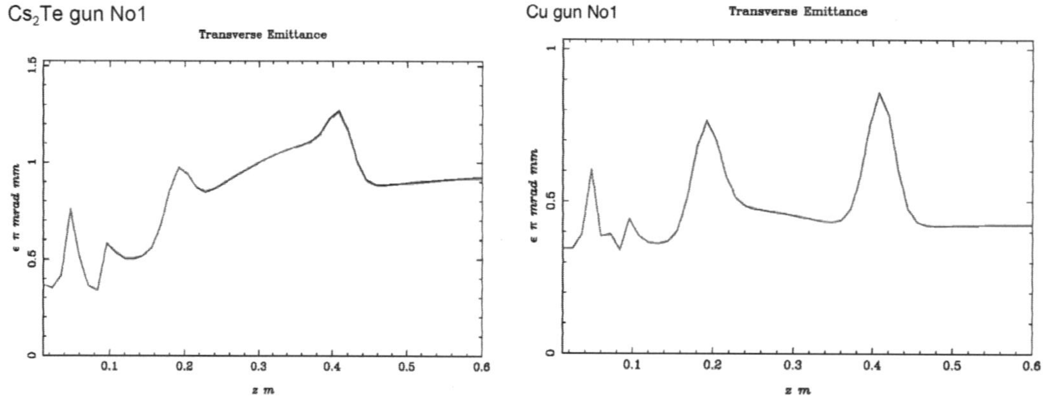

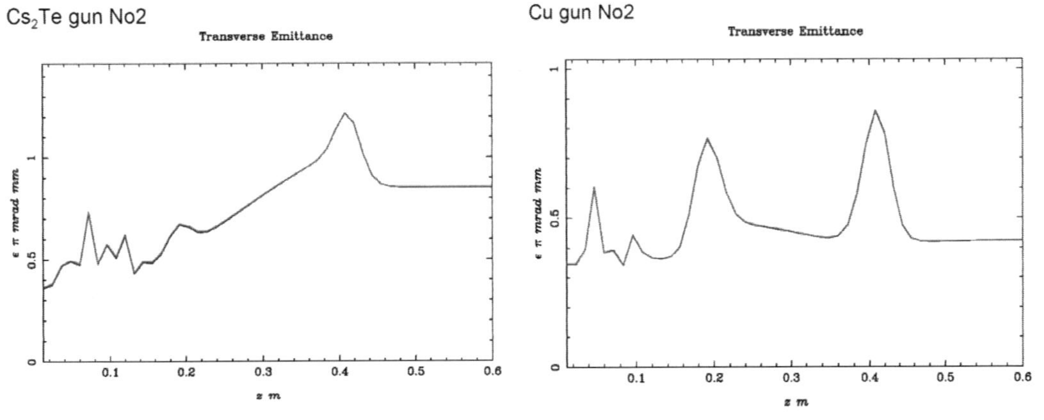

Figure 2: The transverse emittance with respect to the distance from the photocathode is plotted for the two e-guns and the two photocathode materials, respectively.

3.2. The linacs

An S-band (3 GHz) sub-harmonic deflecting structure, operated in the TM110 mode, has been designed to separate the two bunches before the injection into a septum magnet, which separates them into the two FEL lines. The transverse separation between the two beams at the septum is 2.5 mm. The spacing between the two bunches, 6 or 10 X-band RF cycles, is 1.5 or 2.5 RF cycles at S-band. Thus, the two bunches can be placed at the crest and trough of the RF cycle of the sub-harmonic deflector so that the kicks applied to the two bunches are in opposite directions and the separation is maximized for a given kick voltage.

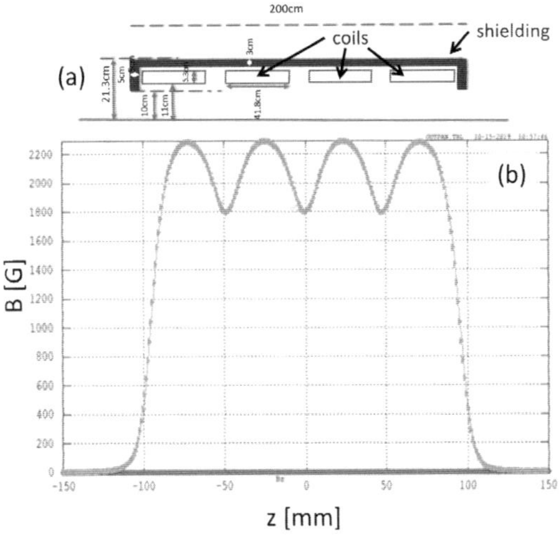

Figure 3: (a) Geometry of the C-band solenoid simulated with Poisson-Superfish; (b) longitudinal magnetic field on axis.

In sequence, a C-band booster is made up of four 2m long traveling wave accelerating structures fed by one klystron and one pulse compressor. The C-band structure parameters are listed in Table 4 and they have been found using the same algorithm illustrated in [11]. Each structure has a linear tapering of the irises. The average dimension of the iris (< a >) and the tapering angle have been chosen in order to maximize the structure efficiency.

Journal of Physics: Conference Series **2375** (2022) 012006 doi:10.1088/1742-6596/2375/1/012006

3.2.1 The C-band structure solenoid

Our contribution is the C-band structure solenoid [12] design for the beam emittance compensation, foreseen around the first accelerating structure. The solenoid has been designed using the code Poisson Superfish [5]. It is a four-coil solenoid with an iron shielding [6], as illustrated in in Figure 3(a), where the simulated geometry is reported. The Figure 3(b) shows the profile of the magnetic field on axis.

Table 4: Main C band structure solenoid parameters

B_{max} [T]	0.22
Bore radius [mm]	100
Total solenoid length [m]	2
Single coil length [mm]	418
Yoke Material Low	Carbon Steel
Total integrated field [Tmm]	40
Good field region radius [mm]	20
Integrated Field variation	10^{-4}
Number of Turns per coil	288
Conductor dimension [mm]	8X8 / bore 6
Nominal Current per coil [A]	278
Nominal Voltage per coil [V]	153
Inductance per coil [mH]	9.17
Resistance per coil [mW]	553
Water Flow Rate per coil [l/min]	2
Temperature drop per coil [C]	20
Pressure drop per coil [bar]	2.43

3.3. The undulators

In the CompactLight design the same undulator line is used in both the SXR and the HXR FEL lines [13]. The undulator parameters have been chosen appropriately to balance the output performance equally between the SXR and the HXR. This feature allows the facility to be more compact and cost-effective. The undulator chain foresees an innovative helical superconducting undulator, with 13mm period and 4.2mm gap followed by an APPLE X afterburner with 19mm period and 5mm gap. The length of each module is 1.75 m.
The main parameters of the SCU and APPLE-X afterburners are summarized in Table 5.

Table 5: Main parameters of the undulators

	Main Undulator	**Afterburner**
Technology	NbTi Superconducting	APPLE-X
Polarisation	Helical	Variable
Period length lw	13mm	~19mm
Maximum aw	1.33	1.80
Beam gap	4.2mm	5mm
Module length	1.755m	1.75m

The following parameters have been considered for the undulators design. The quantitative comparison of the ratio between the FEL peak brilliance and the saturation length as this is a convenient measure of performance vs compactness. Additionally, the chose of the FEL peak brilliance itself, i.e. not normalised to the saturation length, because there is a specific user requirement for a minimum brilliance of 10^{33}ph/s/mm^2/mrad^2/0.1% bandwidth. This allows us an upper limit to be set on the electron beam energy for CompactLight that allows the following criteria to be satisfied:

- The electron beam energy of 5.5 GeV is lower than any other X-ray FEL facility

Journal of Physics: Conference Series **2375** (2022) 012006 doi:10.1088/1742-6596/2375/1/012006

- The photon energy reach of 16 keV is higher than that of SwissFEL which has a beam energy higher than CompactLight
- There is a number of viable options for undulator technology which can be reserved as technology alternatives

4. The 3D CAD model of the baseline layout

The design presented in the CDR includes a facility baseline layout and two main upgrades, with the most advanced option allowing the simultaneous operation of both FEL beamlines, in SXR/HXR pump-probe configuration, at 100 Hz repetition rate. It also includes a preliminary evaluation of the experimental hall, the photon beamlines and the X-ray optics for controlling the focus, intensity and spectral bandwidth of produced photons.
The baseline layout with the tunnel and buildings, of total length is 483 m, is presented as a 3D CAD model [14], an estimation of the conventional facilities, a cost estimate for the overall design and a cost comparison with SwissFEL as a reference facility, complete the design study.
The 3D model is shown in Figure 4a as an ISO view of the CompactLight facility and in the Figure 4b is presented the top view of the total facility, where the shielding blocks have been removed revealing the accelerator.

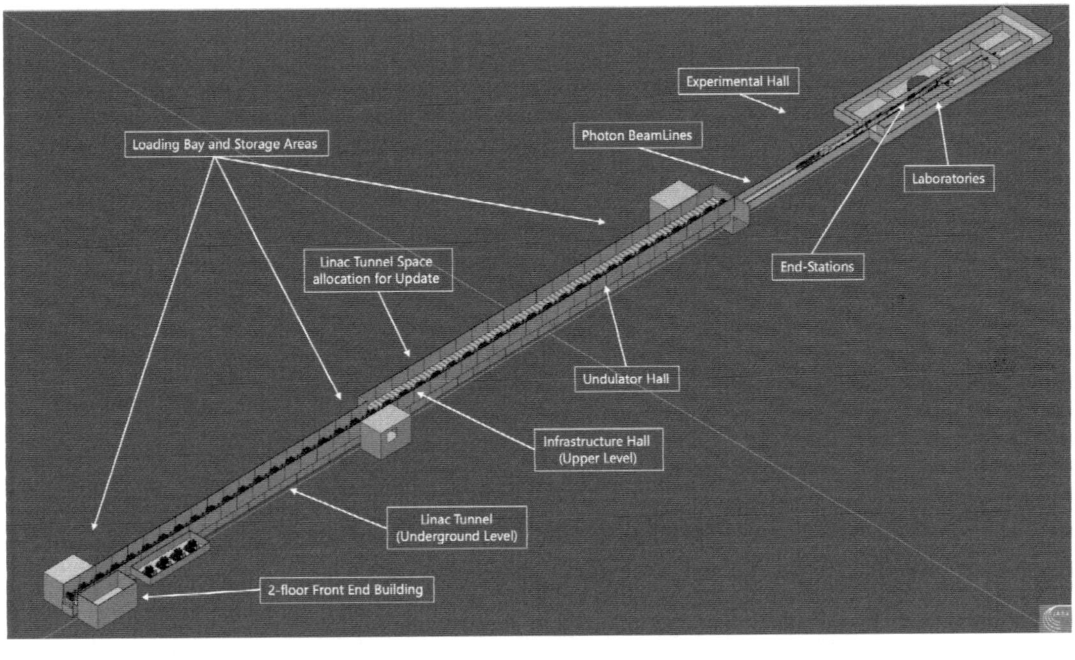

Figure 4a: ISO View of the CompactLight facility with the proposed buildings Naming scheme. Facility building is depicted with the shielding blocks covering the Linac Tunnel.

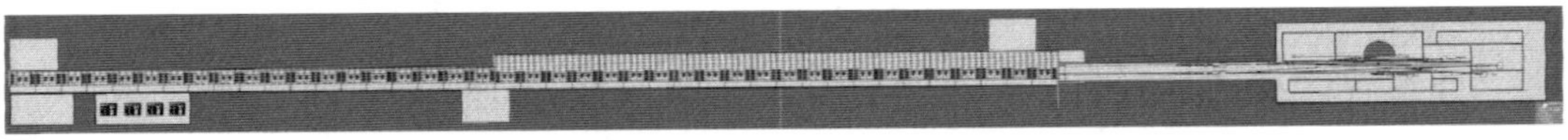

Figure 4b: Top view of the Infrastructure Hall, Linac Tunnel and Undulator Hall with the shielding blocks removed revealing the linear accelerator.

4.1. The 3D CAD model design of the injector

The procedure for the 3D CAD model design is based to the start of design phase up to installation, assembly, commissioning and operation. The design development model being implemented has taken into consideration the initial physics or engineering needs with the requirements of the under-design system in combination with the integration strategy arriving to a preliminary design, simulation and integration. This result via the verification, validation and traceability provides the engineering deliverable of the preliminary design consisting of the machine parts, services, experimental area and the operation integration with the total tunnel hosting the accelerator. In sequence, for the detailed 3D CAD model design many steps should be followed, as: the mock-up prototype fabrication tests and QA/QC check, the series manufacturing the various parts and their assembly, the installation, the commissioning and the operation maintenance. The Figure 5 presents the injector and the consisting parts based on the consolidated injector layout provided by INFN.

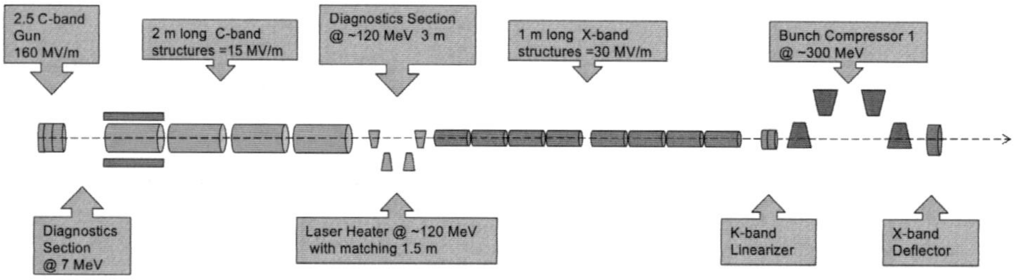

Figure 5: Injector layout

The layout in Figure 5 has been redesigned at the upper part of the Figure 6 with the block symbols foreseen in the parameters list, annex [2], so the 3D CAD model appears at the down part of the Figure 6. The corresponding labels of all the parts of the injector help the connection of the 3D model parts.

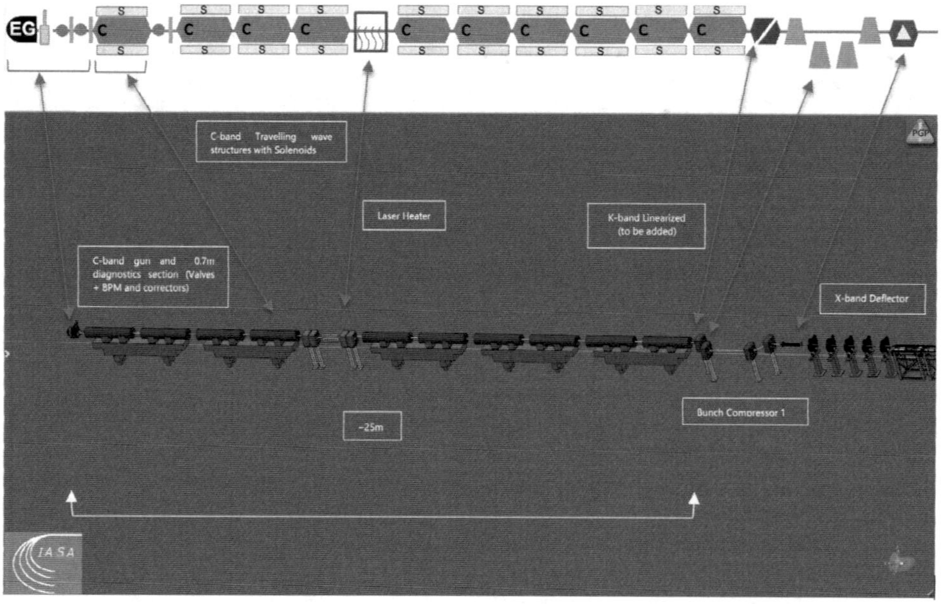

Figure 6: General Layout of the Injector components and the first part of Linac 1

4.2. The Quadrupole magnets

Some additional components, after the Bunch Compressor - BC1, Figure 5, are required to provide the proper parameter optimization like the higher harmonics RF structure for the longitudinal phase space linearization, the X-band RF deflector system to provide the adequate longitudinal diagnostics and the Laser Heater to prevent possible micro-bunch instabilities in BC1 at about 300 MeV beam energy. A 3D CAD model design of series of quadrupole magnets [14], in details of their structure and the power and cooling supply, provide an ideal focusing to the beam in Figure 7.

Figure 7: Detail of the quadrupole magnets design used in CompactLight facility with embedded H and V steerer magnets in a common yoke.

5. The Exploitable Assets

Based on the activities carried out in all the CompactLight Work Packages (WP), a total of 37 main exploitable assets have been produced that can be further utilized in the future, towards the benefit of any current and subsequent stakeholders. In their majority, the identified added value to the industry, potential benefits to the project, and potential benefits to end-users, that stem from the forementioned exploitable assets is estimated to be large to very large.

Those exploitable assets are protected by Copyright have been collected without any utilization of existing Intellectual Property Rights (IPR), except for the intangible pre-existing knowledge and know-how of the involved institutions. However, based on their great value and the advancement of the pre-existing knowledge, they have led to the creation of new IPRs, that are held by the CompactLight Collaboration. Therefore, any further or future use of these assets must respect the relevant licensing regulations, making proper reference to the work to be utilized and the CompactLight project itself. Details of the exploitable assets distribution per WP, on the background (pre-existing before CompactLight), foreground (created by CompactLight), Intellectual Property Rights (IPR) and transfer technology (TT) with regards to each of the forementioned exploitable assets will be published in the recent time [15].

6. The Transfer Technology

An important goal of the CompactLight project is the efficient dissemination and exploitation of the project results through suitable Knowledge and Technology Transfer (KTT) activities to a wider

scientific, technological and industrial society, in order to ensure optimal socio-economic benefits for society from the public R&D investments in the project. This comes in agreement with the concept promoted by the European Competitiveness Council Innovation, KTT is being actively as an essential element of improving European competitiveness, one of the core activities in Europe, where a proactive approach is taken to promote spin-off technologies [16, 17].
The KTT information and cooperation strategy of CompactLight with highly specialized industrial partners provides the mechanism for initial stages of further development and commercialization of the designed accelerator products, which is indispensable in order to proceed on the realization way of our design, thereby creating considerable economic and social impacts from the public funds invested in CompactLight. In addition, the project is increasing the innovation content in the FEL domain, to be viewed as a strategic driver of socio-economic development and impact.

7. Conclusions
The most characteristic, innovative and interesting parts of the CompactLight Collaboration design has been presented; where an enormous, efficient and coherent work by all academic and industrial was achieved to the great result the delivered CDR document, accompanied by decades of other reports with many design details of the FEL parts.

8. Acknowledgement
The authors acknowledge the scientific support by Massimo Ferrario (INFN-Frascati) and his team. They are, also, sufficiently grateful to the CompactLight management for their excellent governance to the project and their special support to our team by the H2020-INFRADEV-777431 CompactLight project funding.

References
[1] Ackermann A *et al* 2007 *Operation of a free-electron laser from the extreme ultraviolet to the water window, Nature Photonics* **1** 336, ISSN: 17494885, DOI: 10.1038/nphoton.2007.76.
[2] CompactLight Collaboration 2022 *The Conceptual Design Report (CDR) of the CompactLight X-ray FEL European Commission under GA No. 777431 XLS Deliverable D3.2*
[3] CLIC Collaboration TDR 2012 *A Multi-TeV Linear Collider based on CLIC Technology* CERN 2012-007
[4] SwissFEL CDR, PSI Bericht Nr. 10-04 July 2010
[5] Warren J L 1987 *Reference manual for the POISSON/SUPERFISH Group of Codes Tech. Rep. LA-UR-87-126*
[6] Gazis N *et al* 2019 *e-Gun and Injector Engineering Proposal ESS Note v 6.4*
[7] Flöttmann K 2016 *ASTRA particle tracking code*
[8] Bacci A *et al* 2016 *GIOTTO: A Generic Code for Demanding Beam dynamics Optimizations Proceedings of the 7th International Particle Accelerator Conference (IPAC 2016), Busan Korea*
[9] Telali E 2019 *Photoinjector study of a novel XFEL in the CompactLight Collaboration MSc thesis National Technical University of Athens*
[10] Tzanetou K 2022 *FEL Photocathode/laser studies MSc thesis National Technical University of Athens*
[11] Diomede M *et al* 2018 *Preliminary RF design of an X-band linac for the EuPRAXIA at SPARC_LAB project, Nuclear Instruments and Methods in Physics Research, Section A: Accelerators, Spectrometers, Detectors and Associated Equipment ISSN: 01689002*
[12] Ferrario M *et al* 2021 *Design of the CompactLight e-gun and injector, with phase space linearizer XLS Deliverable D3.4*
[13] Nguyen F *et al* 2019 *Technologies of the CompactLight Undulator XLS Deliverable D5.1*
[14] Gazis N *et al* 2020 *XLS 3D CAD Model Report ESS Note v 1*
[15] Gazis E *et al* 2022 *The socioeconomic impact of the CompactLight CDR, to be published.*

[16] *Science, Research and Innovation Performance of the EU 2018.pdf*
[17] *Science, Research and Innovation Performance of the EU 2020.pdf*

Development of a muon tomography application with Micromegas detectors

D Amperiadou [1,2], C Petridou [1,2], D Sampsonidis [1,2], S Tzamarias[1,2], K Kordas [1,2], C Lampoudis[1,2], A Leisos [3], A Tsirigotis [3], Ch Tsiafis [4,2]

[1] Department of Physics, Aristotle University of Thessaloniki, University Campus, GR-54124, Thessaloniki, Greece
[2] Center for Interdisciplinary Research and Innovation (CIRI-AUTH), Thessaloniki 57001, Greece
[3] Physics Laboratory, School of Science and Technology, Hellenic Open University, Patras 26222, Greece
[4] Laboratory for Machine Tools and Manufacturing Engineering, Mechanical Engineering Department, Aristotle University of Thessaloniki, Greece

E-mail: damperia@auth.gr

Abstract. The application of muon tomography method in small scale, for the imaging of a five-centimeter side lead cube, is being examined in the present work. The representation of the object is being achieved by using the transmission muography technique, in which the "free-sky" muon flux is compared to the muon flux measured within the acceptance of the detector. This comparison yields information about the absorption of the muons as they pass through the object under investigation. Also, a projective reconstruction method called "Back Projection" is tested by being applied on the data, providing information about the location of the object and its dimensions. This project has been carried out within the frame of EKATY programme, which aims to the innovative imaging of the subsurface of archaeological sites and the interior of structural elements of monuments in three and four dimensions.

1. Introduction

Muon tomography is a method that uses cosmic-ray muons for the imaging of large volumes. There are two types of muon tomography, the muon transmission radiography (muon radiography) and the muon scattering tomography. In the former, the "free-sky" measured muon flux is compared to the muon flux measured within the acceptance of the detector with a potential target located along a certain line of sight, while in the latter the multiple Coulomb scattering of muons is exploited, as they pass through high-Z materials. There are many fields in which muon tomography method has been applied so far. For instance, the first steps have been done in the field of archaeology by Luis W. Alvarez and his team in their search for any unknown chambers in Chephren's pyramid in Egypt [1]. Also, geology, "volcanology" and nuclear safety and security are some fields in which the muon tomography method is applied [2].

Although muon tomography has been mainly used for the imaging of large volumes, this work examines the case of the representation of a small scale object, a five-centimenter side lead cube. Since cosmic ray muons are highly penetrating, the probability of their absorption in this small

Journal of Physics: Conference Series **2375** (2022) 012007 doi:10.1088/1742-6596/2375/1/012007

cube is not significant due to the huge background of high energetic muons. For this reason, a method to reduce this background needs to be applied, described in the next section.

The current analysis has been developed in order to study the muon tomography technique in a small scale experiment. The next step of the EKATY project is the study of the inner structure of the Apollonia's tumulus, an archaeological site near the city of Thessaloniki, applying the muon tomography technique using four stations of Micromegas detectors (Micro-Mesh Gaseous Structure) with an active area of about 0.5 m x 0.5 m. A simulation of the tumulus with potential monuments inside has been developed, as described in [3].

2. The experimental set-up in the laboratory

The experimental set-up is hosted in the Center of Interdisciplinary Research and Innovation of Aristotle University of Thessaloniki. Figure 1 illustrates the whole experimental set-up in the laboratory.

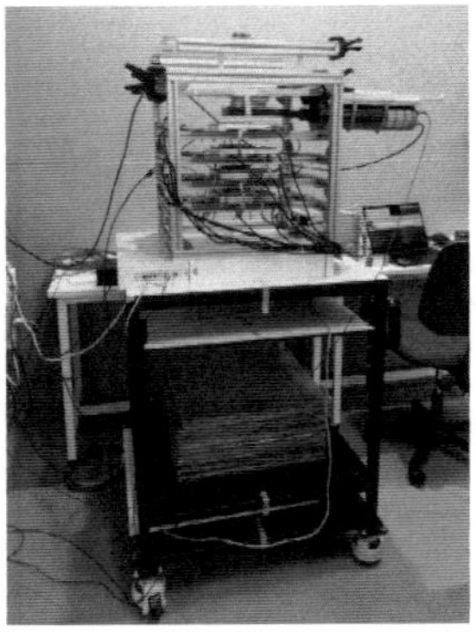

Figure 1: The experimental set-up in the laboratory.

At the top of the layout a cube made of six plates of lead is located along the vertical axis of the telescope. The cube's width is 5 cm and its distance from the top detector was firstly set to 130 mm and later to 230 mm. This is the geometric object on which the muon tomography method will be applied.

The set-up consists of a top scintillator with photo-multiplier tube (PMT). Another scintillator with photodiode is placed 39 cm under the first one. These two scintillators are used for the trigger. Events that give signal on these two scintillators are referred as double coincidence. Events that give signal in one of these two detectors are neglected.

Between these two scintillation-based detectors, four bulk Micromegas detectors [4] are located, for the tracking of incoming muons and they have an active area of 10 cm x 10 cm. The conversion gap is 5 mm and the amplification gap is 0.128 mm . The gas that fills the drift and amplification regions is the $ArCO_2(93{:}7)$. A voltage of 560 V has been applied to the anode, 0 V to the mesh and a voltage of -300 V has been applied to the cathode. The anode 2D readout structure consists of 300 strips along x and y, with a pitch size of 250 μm, providing a high spatial resolution. The distance between two successive chambers is about 7cm.

About 69 cm below the bottom scintillator, there is a third scintillator with photodiode above of which a 20 cm thick block of lead has been placed that can be used as an absorber of muons. The usage of the block of lead with the bottom scintillator is to perform a cut on the muons with high energy focusing on low energy muons which are more likely to be absorbed in the small lead cube. Figure 2 illustrates the energy deposition of passing muons in the 20 cm block of lead from simulated data in GEANT4 [5] along with the initial energy distribution of the input muons in the simulation. The initial energy was chosen to follow a Uniform distribution U(0.1 GeV,4 GeV). The mean value of the energy deposition in the absorber is about 0.246 GeV, meaning that approximately muons with energy lower than this value have been taken into consideration.

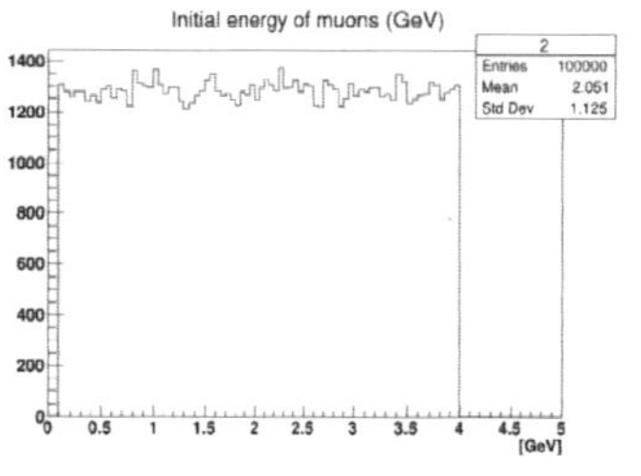

(*a*) Energy of ipnut muons in simulation.

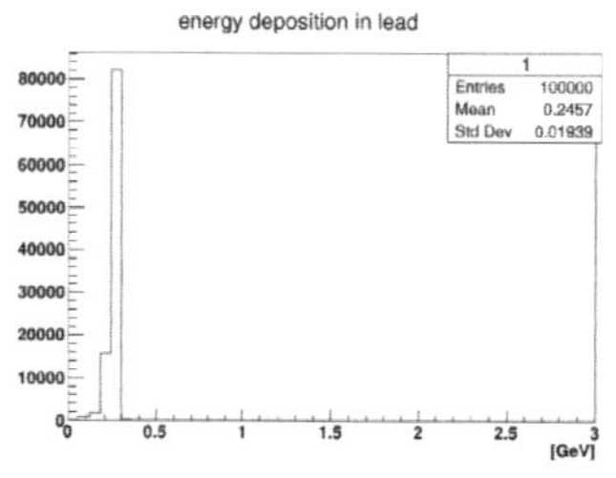

(*b*) Energy deposition of muons in 20 cm lead.

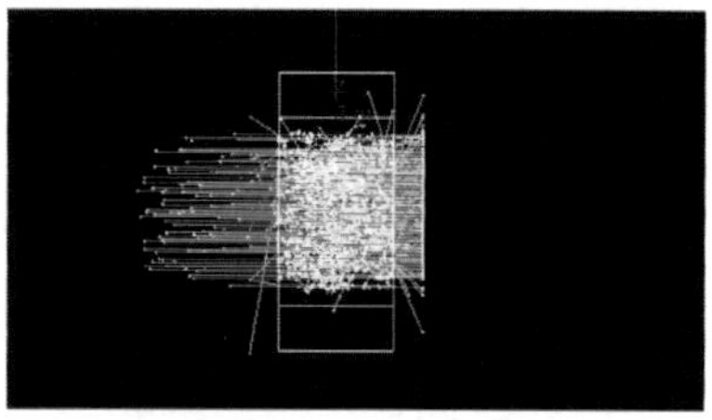

(*c*) Muons passing through 20 cm block of lead.

Figure 2: Initial energy of muons (*a*), energy deposition of muons in 20 cm lead (*b*) and the geometry of simulated muons passing through 20 cm of lead (*c*).

In the first run where the cube was placed at a distance of 130 mm from the top detector, the total days of data acquisition were 12 days, whereas in the second run where the cube was moved 100 mm farther the total days of data acquisition were 27 days. In each run, two data sets were collected, one with the object and one without the object.

3. Track reconstruction

The representation of an object using muon tomography method requires the tracking of the detected muons. The tracking is performed by using the information from the hitted strips in the Micromegas detectors. Since these detectors consist of 2D readout, one track can be reconstructed on xz plane and one on yz plane as well, creating a 2D image with the coordinates of the reconstructed muons' tracks.

For events where one cluster on every layer is present, the extraction of a linear track can be done by directly fitting a linear function on the points that correspond to the weighted averages of the clusters in every layer. However, when there are more than one clusters on a layer the tracking becomes challenging and a tracking method is needed for the reconstruction of a valid linear track from the event. The tracking algorithm, used in these cases, is an iterative algorithm, inspired by the Chain Algorithm described in [6] that gives one track per event, the one with the best χ^2 and is illustrated in the flowchart of Figure 3.

The algorithm begins from the first cluster in the first layer and from the first cluster in the second layer and it performs a linear fit on the weighted averages of these clusters. Next, the

2375 (2022) 012007 doi:10.1088/1742-6596/2375/1/012007

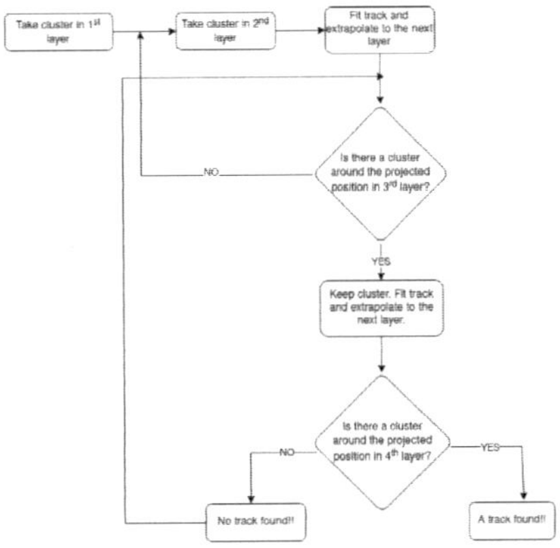

Figure 3: Flowchart of the tracking algorithm.

linear fitted function is extrapolated to the next layer and searches for a suitable cluster in the region around the expected position. The acceptance in which the algorithm can look for a cluster can be set to several values (in strips). In this analysis, this tolerance has been set to 10 strips. If the algorithm finds a cluster on the third layer, it keeps it and performs a new linear fit on the three points now. In the next step, the algorithm searches again for a suitable cluster on the fourth layer. If it finds one, then a track is reconstructed. If the algorithm does not find a cluster within the search region, it goes back to the previous layer to try with another cluster considering all the possible combinations and at the end it may return no track (if it does not find a cluster on a layer after trying all the combinations), one track or more than one track. In the case where more than one track is returned, the one with the smallest χ^2 is kept, as the best track.

In this analysis, events with signal in all four detectors and with less than five clusters in every layer are considered. In the first run, the total events from the data set with the cube were 38305 and the total reconstructed tracks by the tracking algorithm were 28322. The data set without the cube had 47700 events with signal in all detectors and 34689 tracks were reconstructed by the tracking algorithm. In the second run in the data set with the cube, the total events were 90369 and the reconstructed tracks were 60424. In the data set without the cube, the total events were 102610 and the tracking algorithm reconstructed successfully 69183 tracks. These data come from double coincidence events with VETO from the third scintillator.

Also, a data set with triple coincidence events has been collected in order to examine the acceptance of the three scintillation detectors by the distributions of the incident angles of the reconstructed tracks. Figure 4 shows the results from the analyzed data and the results from simulated data in GEANT4 for validation. In both cases the acceptance of the three scintillation detectors is in the range of $75^o - 105^o$ incident angle, although the distributions are not fully compatible at angles larger or smaller than 90°.

In the analysis described in the next section, for the application of muon tomography method, tracks with slope in the interval of the three scintillators' acceptance have been taken into consideration obtained from double coincidence events with VETO.

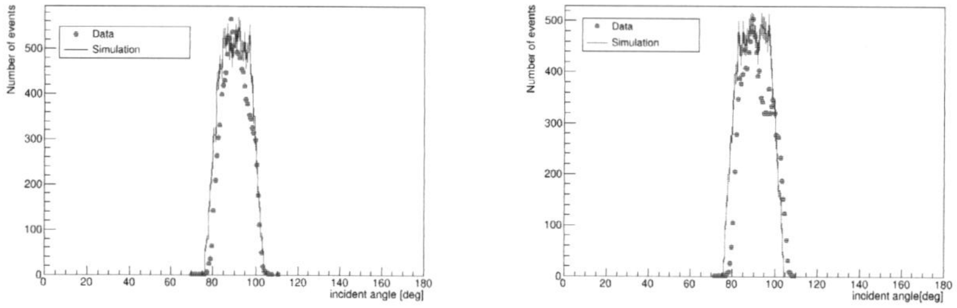

(a) Incident angles of reconstructed tracks from data and simulation on xz plane.

(b) Incident angles of reconstructed tracks from data and simulation on yz plane.

Figure 4: Distribution of incident angles on xz and yz plane from triple coincidence events .

4. Back projection method

The back projection method, described in [7], is an ideal method for the imaging of objects which are surrounded by material of different density. Also, in order for this method to be effective the detector's dimensions should not be negligible with respect to the object's ones. Under these assumptions, the approximate distance of the object from the top detector and its dimensions can be determined.

The basic idea of this method is the back projection of the reconstructed muons' tracks to parallel planes which are also parallel to the top detector's plane. For each back projection plane a bin size (pitch) is defined which scales linearly with the distance of the plane from the top detector,

$$p(x) = p_0 + \delta_\theta x, \tag{1}$$

where p_0 is the spatial resolution of the single tracking plane, δ_θ is the angular resolution of the telescope, defined as the ratio of the spatial resolution to the distance between the first and last tracking detectors and x is the distance between the back projection plane and the top Micromegas. The back projection planes, selected at several distances from the top detector, are two-dimensional histograms filled with the ratio of the projected muon's position to the corresponding pitch. Such histograms are created from the data set with and without the object in each run. Next, for all the positions, we get the subtraction of the histogram with the object from the histogram without the object and each bin is also divided by the sum of the bin contents of the two histograms. The RMS of this new histogram corresponds to the width-to-pitch ratio and reaches a minimum value at the approximate distance where the object is located. Also, according to this method, the Full Width at Half Maximum (FWHM) of the histograms after the subtraction at the estimated distance yields the object's dimensions. In this analysis instead of FWHM proposed in [7] the 2σ parameter used for the determination of the object's width.

In order to specify the pitch for each projection plane according to relation (1), the spatial resolution p_0 should be defined, which will also determine the angular resolution δ_θ. For this purpose, a simulation model of the single Micromegas detector has been developed in Garfield++ software toolkit [8] and the distribution of the residuals between the weighted average of the clusters and the real hit position has been generated for different incident angles of the passing muons. In the simulation, the starting point of the particle and its momentum were chosen from a Uniform distribution U(0.1 cm,0.4 cm) and U(1 GeV/c,4 GeV/c), respectively. These specific results have been produced using the Aristotle University of Thessaloniki (AUTh) High Performance Computing Infrastructure and Resources.

The results in the figure below shows the worsening of the spatial resolution with the incident

Journal of Physics: Conference Series **2375** (2022) 012007 doi:10.1088/1742-6596/2375/1/012007

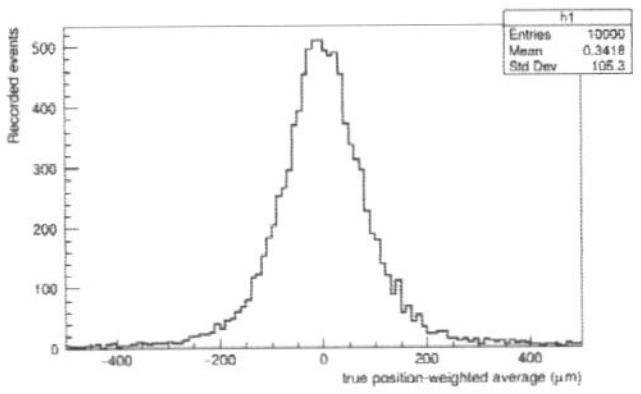

(a) Incident angle=90°.

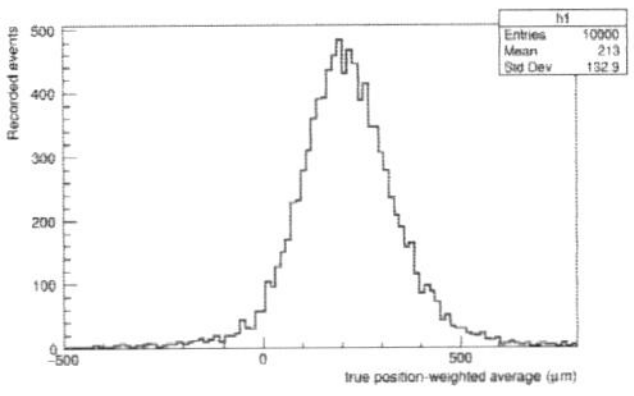

(b) Incident angle=95°.

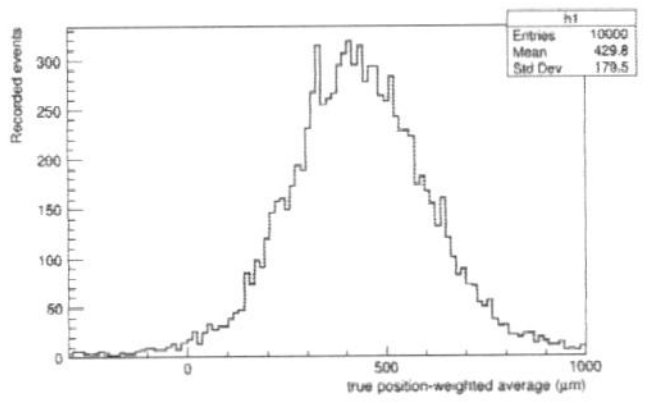

(c) Incident angle=100° .

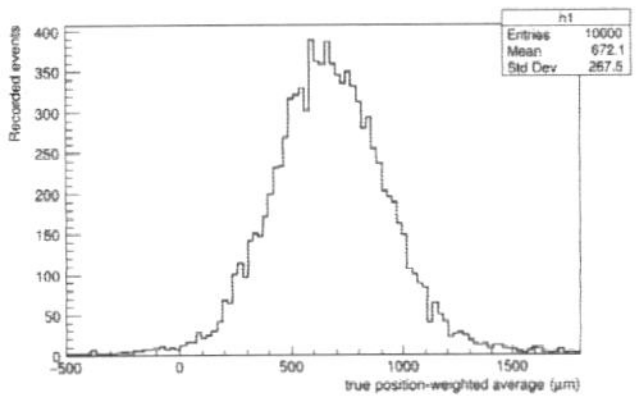

(d) Incident angle=105 ^{o}.

Figure 5: Distribution of the residuals (true position-weighted average) for four different incident track angles .

angle. Also, the residuals do not have mean value of zero as we move to different incident angles than 90°. This is due to the cluster centroid method which gives the crossing point systematically higher due to larger (extended) cluster. In the analysis, since tracks with incident angles in the interval of the three scintillators' acceptance are considered, the "worst case scenario" has been chosen, where the incident angle deviates 15° from the vertical and the spatial resolution is 267.5 μm. This spatial resolution yields an angular resolution of $\delta\vartheta$=0.0012 rad.

In the following figures, the results from the application of the back-projection method in the first and second run are depicted. Figure 6 illustrates four indicative back projection planes at four different distances from the top detector, after the subtraction. The two-dimensional histograms are filled with the ratio of the x or y coordinate to the corresponding pitch. The pitch value of each projection plane is calculated using relation (1). The behaviour of the two-dimensional histograms is the expected one, where a focused region with an excess of flux is distinguished as we move closer to the object. The real distance between the cube and the top detector is 130 mm. The displacement of the histograms as we move farther may be due to imperfect alignment of the telescope, with respect to the vertical axis.

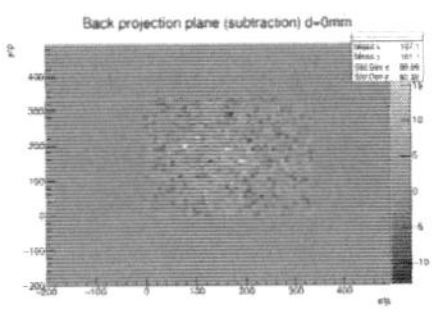

(a) Back projection plane at d=0 mm.

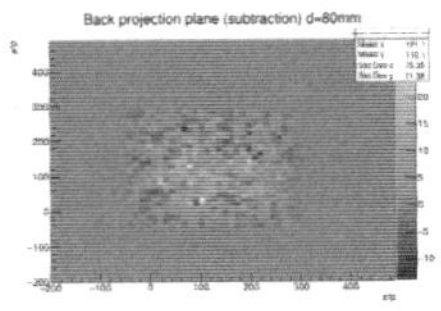

(b) Back projection plane at d=80 mm.

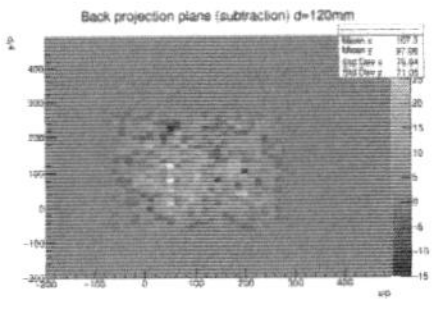

(c) Back projection plane at d=120 mm.

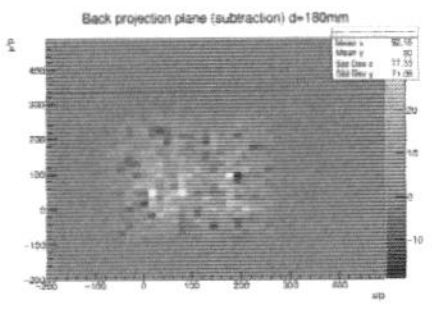

(d) Back projection plane at d=210 mm.

Figure 6: Back projection planes at four different positions (The real distance of the cube from the top detector is 130 mm).

In figure 7 the width-to-pitch ratio, which corresponds to the standard deviation of the projections on xz and yz planes, is plotted as a function of the distance from the top detector. The minimum for both graphs has been derived by fitting a second order polynomial to the data in a small interval around the minimum. The errors correspond to the errors of the standard deviation and the error in the estimated distance is obtained from error propagation taking into consideration the parameters' errors of the fitted function and their correlation from the correlation matrix.

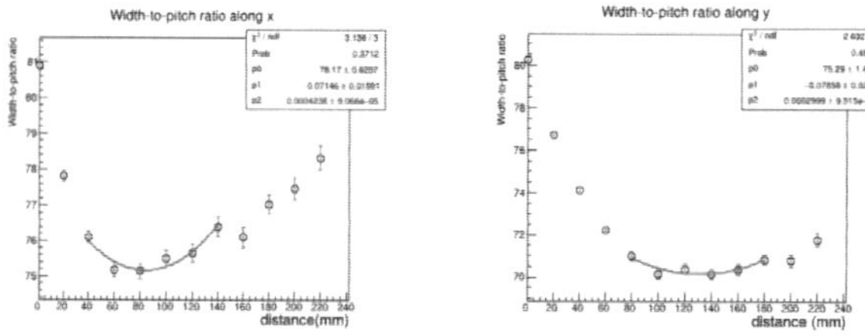

(*a*) The width-to-pitch ratio along x axis.

(*b*) The width-to-pitch ratio along y axis.

Figure 7: The width-to-pitch ratio along x (*a*) and along y (*b*) axis as a function of the distance of the projection plane from the top detector.

The results from the fitted functions are,

$$\begin{aligned} D^x_{min} &= 84.32 \pm 3.16mm \\ D^y_{min} &= 131 \pm 4.97mm. \end{aligned} \tag{2}$$

The estimated distance of the object from the detector seems to be underestimated along x axis, whereas the result along y axis is very close to the real distance of 130 mm. Also, for the estimation of the cube's dimensions, the plot of 2σ along x and y axis is provided and it is fitted with a linear function in the same interval as in the case of the width-to-pitch-ratio.

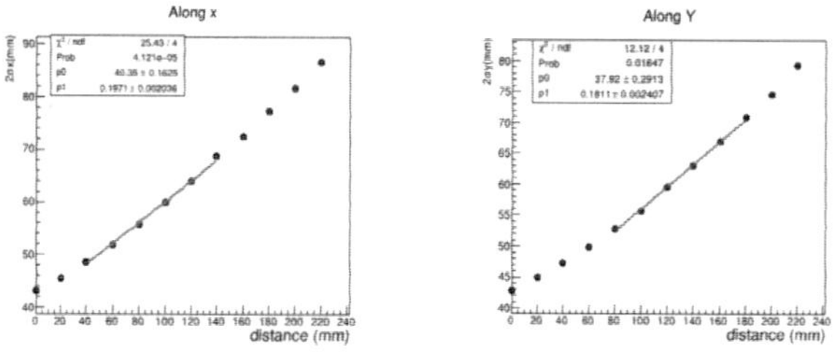

(*a*) 2σ along x axis.

(*b*) 2σ along y axis.

Figure 8: The 2σ parameter along x (*a*) and y (*b*) axis as a function of the distance of the projection plane from the top detector.

The values of the linear functions at the distances determined before are,

$$\begin{aligned} 2\sigma^x(84.32mm) &= 56.97 \pm 1.4mm \\ 2\sigma^y(131mm) &= 61.64 \pm 1.30mm \end{aligned} \tag{3}$$

and they are approximately compatible with the real width of the cube, which is 50 mm.

Also, the same analysis has been carried out for the second run, where the cube was placed at a distance of 230 mm from the top Micromegas. Figure 9 depicts four indicative back projection planes after the subtraction at four different positions.

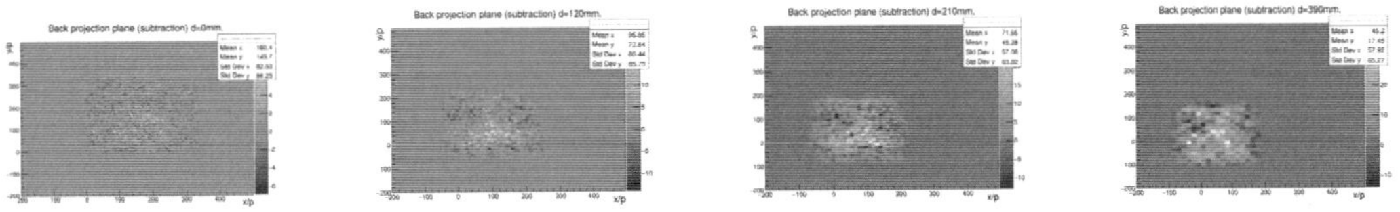

(*a*) Back projection plane at d=0 mm. (*b*) Back projection plane at d=120 mm. (*c*) Back projection plane at d=210 mm. (*d*) Back projection plane at d=390 mm.

Figure 9: Back projection planes at four different positions (The real distance of the cube from the top detector is 230 mm).

As we move farther from the top detector and closer to the object, the image becomes sharper and an excess of flux near the centre shows up . Since the images are obtained from subtraction of the two histograms (with and without the object) there are bins containing negative counts in some regions which are due to statistical fluctuations.

In figure 10 the width-to-pitch ratio is plotted as a function of the distance between each projection plane and the top detector. The graph is fitted with a second order polynomial in the region around the minimum. The distance at which the fitted function reaches the minimum value corresponds to the estimated distance between the cube and the top detector.

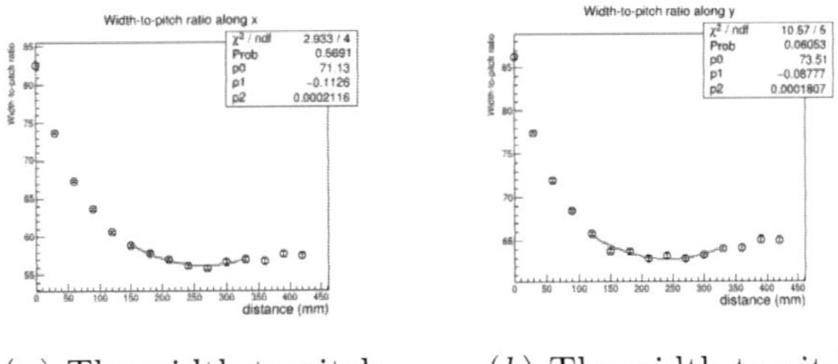

(*a*) The width-to-pitch ratio along x axis. (*b*) The width-to-pitch ratio along y axis.

Figure 10: The width-to-pitch ratio along x (*a*) and y (*b*) axis as a function of the distance of the projection plane from the top detector.

The results from the fitted functions are,

$$\begin{aligned} D^x_{min} &= 266.06 \pm 6.62mm \\ D^y_{min} &= 242.86 \pm 5.69mm. \end{aligned} \tag{4}$$

,which are close to the real distance of 230 mm. Also, the 2σ parameter is plotted as a function of the distance in order to determine the cube's width as shown in figure 11.

The values of the linear functions at the distances determined before are,

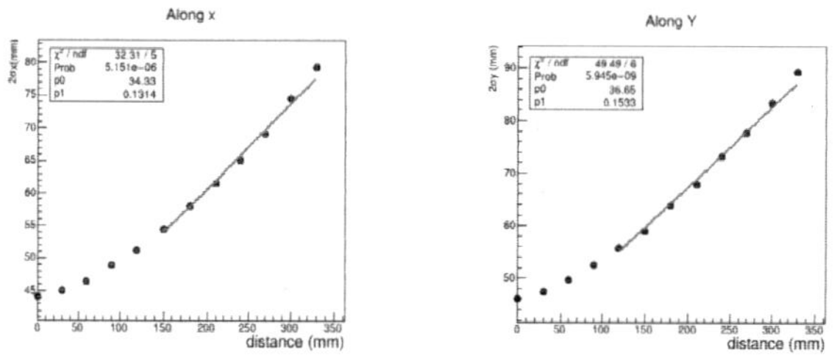

(a) 2σ along x axis. (b) 2σ along y axis.

Figure 11: The 2σ parameter along x (a) and y (b) axis as a function of the distance of the projection plane from the top detector.

$$\begin{aligned} 2\sigma^x(266.06mm) &= 69.30 \pm 2.30mm \\ 2\sigma^y(242.86mm) &= 73.87 \pm 2.62mm. \end{aligned} \tag{5}$$

and they seem to overestimate the real width.

5. Conclusions and prospects

In this work the muon tomography method is applied in a small scale experiment, for the representation of a five-centimeter side cube of lead. Although muon transmission radiography can be applied for the imaging of small objects, a technique to reduce the high background from the high energetic and penetrating muons, should be used. In our case, the VETO operation is applied which excludes muons that can penetrate a 20 cm block of lead. In addition, back projection method is tested, an ideal method for the imaging of objects of high or low density with respect to the material's surround it. The results of the current study show that this method can be used in this experiment for the estimation of the approximate distance between the investigated object and the top detector. However, the object's dimensions determined by the 2σ parameter seem to be overestimated. Also, although the Micromegas chambers have the same pitch (250 μm) in X and Y, the strip width is different, namely 150 μm in X and 75 μm in Y in order to have the same capacitance. These are bulk Micromegas chambers, where the strips along y are above the strips along x and therefore closer to the cathode. This may lead to different results along the two dimensions.

As far as the future work is concerned, the method can be tested for different objects and for different distances, too. Apart from the muon transmission radiography, the muon scattering tomography could be also applied, which may be more ideal for the imaging of small objects. In that case, two Micromegas detectors should be placed above the object and other two below and the scattering angle of the passing muons could be determined, due to their multiple Coulomb scattering as they pass through the material. Also, the tracking method can be optimized. For instance, another algorithm can be used which will be able to extract more than one tracks from an event. Last but not least, within the frame of EKATY program, new Micromegas detectors of about 0.5 m x 0.5 m are under construction and will be set up in the next months for the application of muon tomography to archaeological sites.

6. Acknowledgements

This work was carried out in the framework of the research project EKATY (Innovative imaging of the subsurface of archaeological sites and the interior of structural elements of monuments in 3 and 4 Dimensions). The project is under the code T6ΥΒΠ-00211 in the framework of the Operational Programme Competitiveness, Entrepreneurship and Innovation 2014-2020 (EPAnEK) funded after the decision 97476/I2/18-06-2019 (ΑΔΑ: ΩΕ1Ω4653ΠΣ-ΒΑΓ) of the Greek General Secretariat for Research and Technology.

References

[1] Alvarez L W, Anderson J A, Bedwei F E, Burkhard J, Fakhry A, Girgis A, Goneid A, Hassan F, Iverson D, Lynch G, et al. 1970 Search for hidden chambers in the Pyramids *Science* **167** 832–9

[2] Bonechi L, D'Alessandro R and Giammanco A 2020 Atmospheric muons as an imaging tool *Reviews in Physics* **5** 100038

[3] Kalaitzidou I, Tsiledakis G, Lampoudis C, Petridou C, Sampsonidis D, Kordas K, Tzamarias S, Leisos A, Tsirigotis A and Tsiafis C 2021 Feasibility Study of Muon Tomography application in a non-invasive representation of tumulus *Journal of Physics: Conference Series* **2105** 012017

[4] Giomataris I, De Oliveira R, Andriamonje S, Aune S, Charpak G, Colas P, et al. Micromegas in a bulk. Nuclear Instruments and Methods in Physics Research Section A: Accelerators, Spectrometers, Detectors and Associated Equipment. 2006;560(2):405–8.

[5] https://geant4.web.cern.ch/

[6] Bortfeldt J. The Floating Strip Micromegas Detector. Jan. 2015. doi: 10.1007/978-3-319-18893-5

[7] Bonechi L, D'Alessandro R, Mori N and Viliani L 2015 A projective reconstruction method of underground or hidden structures using atmospheric muon absorption data *Journal of Instrumentation* **10**

[8] https://garfieldpp.web.cern.ch/garfieldpp/

Air shower radio signal electric field orientation as measured with the Astroneu Cosmic Ray telescope.

S Nonis[1], A Leisos[1], A Tsirigotis[1], G Bourlis[1], K Papageorgiou[2], I Gkialas[2], I Manthos[3] and S Tzamarias[4]

[1] Physics Laboratory, School of Science and Technology, Hellenic Open University, Patras 26222, Greece

[2] School of Engineering, Department of Financial and Management Engineering, University of the Aegean, Chios 82100, Greece

[3] School of Physics and Astronomy, University of Birmingham, B15 2TT, United Kingdom

[4] Department of Physics, Aristotle University of Thessaloniki, Thessaloniki 54124, Greece

E-mail: snonis@eap.gr

Abstract. The Astroneu cosmic ray telescope is a small scale hybrid array consisting of both scintillator counters and electromagnetic radiation detectors in the radio-wave frequency (RF) band (antennas). Astroneu was installed and operated in the area of the Hellenic Open University (HOU) campus near the city of Patras in Greece. In the present development phase, the Astroneu telescope includes two stations consisting of 3 scintillation detectors modules (SDM) and one RF antenna while a third station includes 3 particle detectors and 4 RF antennas (3SDM-4RF). In this context we present the resent results from both the 3SDM-4RF autonomous station and simulations related to the estimation of the direction of the electric field emitted during the air shower development. The electric field measured in the ground level is the superposition of the two dominant emission mechanism, the time depended transverse current induced by the geomagnetic field and the net negative charge variation at the shower front. Since the electric field emitted by the two contributions is polarized in different directions the measured electric field in the ground encloses information about the charge-excess to geomagnetic ratio (CGR). Furthermore the orientation of the electric field emitted by charge-excess mechanism is strongly depends from the distance to the shower core (the intersection of the shower axis with the ground level). In this study we use the core information as reconstructed using the radio data and simulations. The estimated charge-excess to geomagnetic ratio is in agreement with previous studies which reveals that the shower core reconstruction method is efficient. Finally we report on that CGR measurements can be used for an efficient noise rejection in a future self trigger mode

Keywords: Cosmic rays, Astroneu, radio detection of extensive air showers, electric field orientation.

1. Introduction

The term cosmic rays (CR) is commonly used to describe very high-energy atomic nuclei that bombard the Earth from extraterrestrial sources. When these particles enter the Earth's atmosphere they create extensive air showers (EAS) of secondary particles. The detailed study of EAS is a significant tool for deriving the properties of the primary CR such as its energy, its incoming direction and its composition. The relativistic motions of secondary electrons and positrons during the EAS development produce coherent electromagnetic pulses in the MHz frequency domain [1]. This radio frequency (RF) radiation can be detected by antennas as successfully reported by a series of experiments over the last two decades [2, 3, 4, 5, 6]. Because of the significant theoretical and experimental progress of the recent years RF detection is recognized as an efficient technique for detecting EAS and reconstructing the main CR's parameters (see [7, 8, 9, 10] for summary)

Two physical processes contribute essentially to the RF emission during the EAS development. Electrons and positrons while traveling through the geomagnetic field undergo a systematic deviation in opposite directions under the influence of the Lorentz force ($\vec{F}_L = e(\vec{v}_{sh} \times \vec{B}_{geo})$) which is perpendicular to both the geomagnetic field ($\vec{B}_{geo}$) and to the EAS's direction of propagation ($\vec{v}_{sh}$) [11]. The electric field emitted by the geomagnetic mechanism is oriented in the direction of the Lorentz force while its strength is proportional to $\sin a$, with a the angle between $\vec{B}_{geo}$ and $\vec{v}_{sh}$ (geomagnetic angle). A second contribution to the RF signal comes from the excess of electrons in the shower front as expected due to medium ionization and positron annihilation [12]. The corresponding electric field is radially polarized to the shower axis which means that, unlike the geomagnetic contribution, the electric field of the charge-excess mechanism depend on the location of the observer.

The Astroneu is a combined CR telescope consisting of both scintillator counters and RF antenna detectors [13]. The array is placed in the campus of the Hellenic Open University (HOU) on the outskirts of the city of Patras in Greece. During the 8 years of operation of the telescope, studies have been carried out which are related to detection and reconstruction of EAS parameters using both the particle and the RF detectors [13, 14]. Despite the fact that the telescope operates in an area with high levels of electromagnetic noise, the RF signal emitted during the EAS development was successfully distinguished from the noise by considering the appropriate selection criteria [15]. Studies on the timing and amplitude characteristics of RF signals were performed by combining RF and SDM data as well as simulation predictions [16, 17]. Moreover the response for the RF system was established using the antenna's effective length and the measured RF spectrum in order to evaluate the EAS's axis direction [10, 18].

Here we present a method for determining the EAS core using the RF signal and Monte Carlo (MC) simulations, together with measurements of the direction of the electric field emitted during the EAS expansion, which lead to the Charge excess to Geomagnetic Ratio (CGR) computation. The instrumental setup is briefly reported in section 2. The EAS core reconstruction technique is described in section 3 in combination with simulations results. Section 4 presents the data sample while in section 5 is described the CGR calculations. Results from CGR application to both data and simulation sample are presented in section 6. Section 7 summarizes the conclusions.

2. The Astroneu infrastructure

The detection system applied at the Astroneu telescope incorporates scintillator detectors and RF antennas, grouped in three independent stations (A, B and C). In its current operating phase stations B and C include 3 Scintillator Detector Modules (SDM) [14] and one RF antenna [19] (3SDM-1RF) covering the frequency range 1 - 200 MHz. Station A consist of 3 SDMs and 4 RF detectors as depicted in figure 1-right. Part of each station's equipment are electronic systems for triggering, digitization and Data Acquisition (DAQ) together with a GPS-based timing system.

Each SDM is made of 160 scintillating tiles mounted on a surface of 1 m^2. The light producing from the interaction of the EAS particles with the scintillation material is led through optical fibers to the Photomultiplier tube (PMT) and then converted into a voltage pulse. The 3 SMDs waveforms are collected and digitized by a Quarknet electronics board [21], while timestamping is provided by a GPS receiver. A detail report of the SDMs operation is given in [22]. When the signals from the 3 SDMs exceed a predefined threshold a trigger pulse is sent to the antenna's electronics. A schematic illustration of the station's A connections is shown in figure 1 (left).

The RF antenna of each station is formed by two bow tie shaped dipoles which are aligned

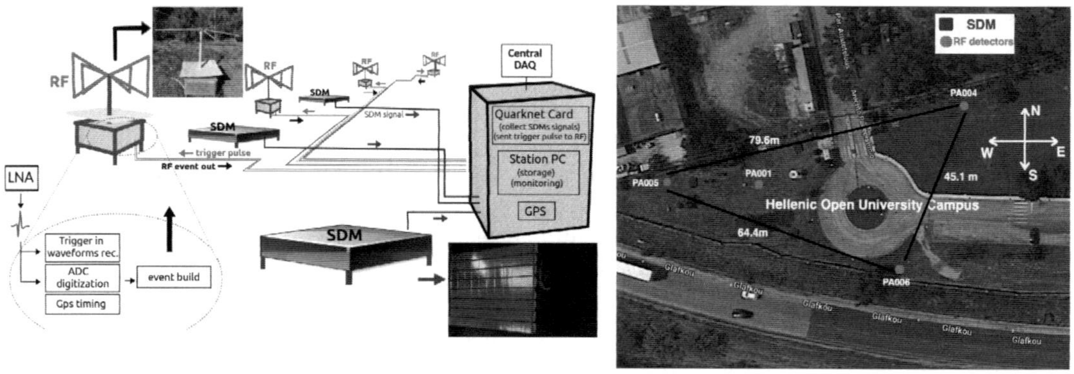

Figure 1. (*Left*) A schematic diagram of the connections in the 3SDM-4RF station. (*Right*) The detector systems of the 3SDM-4RF [20] station in HOU campus. The blue squares denote the 3 SDMs while the red circles the 4 antennas.

along the directions of East-West (EW) and North-South (NS). Furthermore the RF system is equipped with electronics for signal amplification, digitization and event time-stamping (GPS card) [19]. The RF signal read out for both EW and NS polarization direction is activated by an external trigger from the 3 SDMs. The waveforms were sampled at a rate of 1 GS/s (2560 points for a time period of 2.56 μs) and timestamped from the antenna's GPS card. Measurements from both the SDMs and the RF detectors of the station are stored to the Global Data and Control Server (GDCS), where each event is reconstructed offline using the GPS time-tags.

The 3SDM-4RF station of the Astroneu array has the potential of reconstructing EAS of energy more than 10 TeV with an angular resolution of 3.5 degrees at a rate of 17.5 per hour. The efficiency of the Astroneu array in detecting and reconstructing EAS using the data from the SDMs of one station (single station operation) or by combining the data information from two stations (multiple station operation) is reported in [22, 23].

3. Method for EAS core reconstruction - Results from simulations

3.1. The method

An essential step towards to CGR computation is the reconstruction of the EAS core which means the point of intersection of the EAS axis with the ground plane. The RF signal and simulations are used to determine it. For an experimental event of known arrival direction (zenith angle θ_{sh} and azimuth angle ϕ_{sh}) we run a set of 60 simulations with SELFAS package [24] which calculate the electric field in 250 points of a virtual array as depicted in figure 2-left. The center of the virtual array is identified with the center of the station A. All the simulations correspond to the same primary energy (arbitrary set to 10^{18} eV for induced RF signals well above the background noise), core position (the center of the station) and direction (the direction of the experimental event). Since the main CR candidates at ultra high energies are protons (p) and iron nuclei (Fe), 40 simulations correspond to proton primary while 20 of them to iron nuclei covering a reasonable range of X_{max} according to the fluctuations that arise between EAS of the same energy. Due to the fact that interaction cross section with air nuclei lead to larger X_{max} fluctuations in case of proton we run more simulations with proton primary. From the average of the simulated electric fields and adding the background noise (as measured in previous studies [10, 17]) we make the map of the electric field in the ground. The field at positions where there are no virtual antennas is calculated by linear interpolation. The next step is to move the core position (and thus the map of the electric field in the ground) to a circular area of radius 200 m

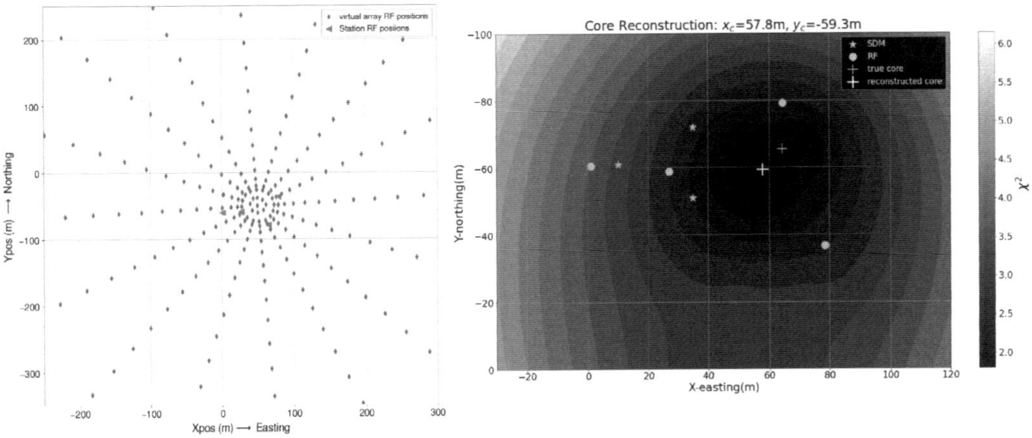

Figure 2. (*Left*) The positions of the virtual antennas used for the simulations (blue diamonds) and the positions of the station's antennas (red triangles). (*Right*) χ^2 density map calculated for a simulated event with known core (green cross). The reconstructed core corresponds to the white cross.

around the center of the station (in order to reduce computational time) and calculate for each new core position the quantity χ^2 given by the formula

$$\chi^2 = \sum_{i=1}^{N_{ant}=4} \left(E_{data}(x_{ant_i}, y_{ant_i}) - E_{sim}(x_{ant_i}, y_{ant_i})\right)^2 / \sigma_i^2 \quad (1)$$

where E_{data}, E_{sim} is the electric field as measured by the experimental data and the (60) simulations respectively at the positions of the 4 antennas of the station. For σ_i we replace the standard deviation of the electromagnetic background that has been measured in each antenna. The point (X_{core}, Y_{core}) at which the quantity χ^2 is minimized gives us the position of the core. Figure 2-right shows an example of χ^2 minimization for a simulation event of which the true core is known (green cross).

3.2. Application of the method to simulation data

The method for the core reconstruction was initially applied to a sample of simulation events. Generating simulation events is a 5-step process. First, we produced high energy shower events (in the energy range $10^{17} - 2 \cdot 10^{18}$ eV) with the CORSIKA simulation package [25]. A detailed description of the EAS parameters and the used interaction models can be found in refs [17, 18]. The second stage is the RF signal development with SELFAS package, which calculates the electric field of the RF signal in the antenna's positions. The electric field is convoluted with the antenna response and the generated signal is finally distorted with noise as calculated from a large number of background events for the antennas of the station. In the final step the same event selection criteria are imposed as those for the experimental data [1]. From the simulation data were selected those whose the true core is distributed in a radius of 200 meters around the center of the station. The final sample consist of 20,500 events. For each simulation event the position of the core is reconstructed with the method described previously and compared to its true position. In figure 3 is depicted the distributions of the difference between the true

[1] In the standard event analysis after filtering the RF signal in frequency range 30-80 MHz cut values for the signal to noise ratio (≥ 6), the pulse rising time (≤ 28 ns) and the degree of polarization (≥ 0.85) are implemented. For details see [15, 17, 18].

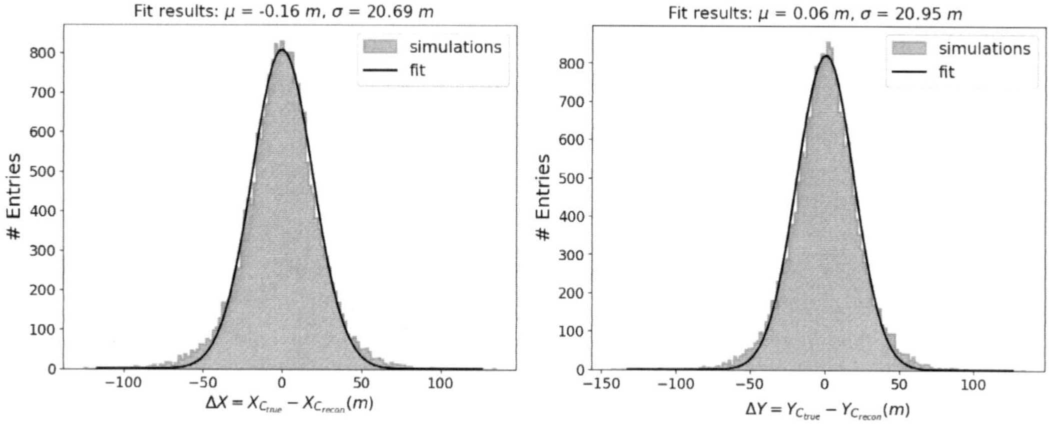

Figure 3. (*Left*) The distribution of the difference $X_{core_{true}} - X_{core_{recon}}$ between the true and the reconstructed X coordinate of the core position. (*Right*) The distribution of the difference $Y_{core_{true}} - Y_{core_{recon}}$ between the true and the reconstructed Y coordinate of the core position.

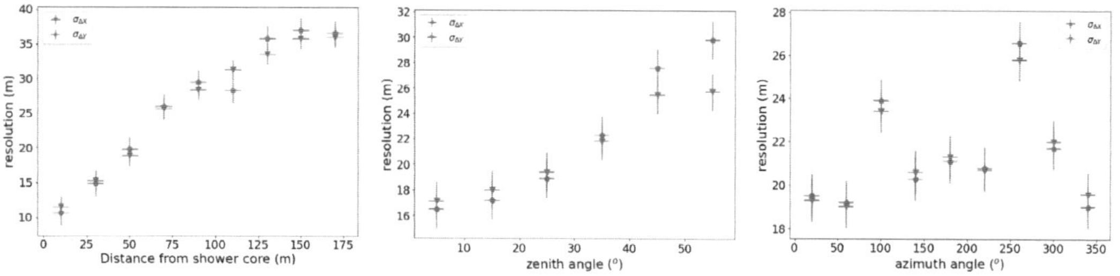

Figure 4. The X core coordinate (blue points) and the Y core coordinate (red points) resolution as a function of the distance to shower core (left), the zenith (middle) and azimuth angle (right). The resolutions are evaluated as the sigmas of the Gaussian functions that fit the distributions of the differences $X_{core_{true}} - X_{core_{recon}}$ and $Y_{core_{true}} - Y_{core_{recon}}$ respectively.

and the reconstructed X (left) and Y (right) core coordinates. Both distributions is fitted with Gaussians functions with mean around zero and sigma equal to 20.69 m for the X coordinate and 20.95 m for Y coordinate. In figure 4 is shown the resolution in estimating the X (blue points) and Y core coordinate (the red points) as function of the distance to shower core (left), the zenith (center) and azimuth (right) shower angle. The resolution in all cases is evaluated as the sigma of the gaussian that fits the distribution of the differences $X_{core_{true}} - X_{core_{recon}}$ and $Y_{core_{true}} - Y_{core_{recon}}$ respectively. Resolution worsens with increasing distance from the core and zenith angle as expected [2] while it is approximately constant as a function of azimuth angle except the north-south direction (90 and 270 degrees) in which the dominant geomagnetic mechanism predicts small amplitudes. Furthermore, since the antenna response is reduced to large zenith angles inclined showers will induce weaker signals on the antennas, relative to the signals induced by vertical showers.

[2] Since the core reconstruction method depends on the strength of the RF signal the resolution is expected to decrease with increasing distance and zenith angle as weaker signals induced in these cases on the antenna's dipoles.

4. The data sample

The sample of experimental data used in this analysis were collected from the 3SDM-4RF station during a 4-year operating period (approximately 25,900 hours). In the experimental data collected from station A and matching the event selection criteria, the EAS axis direction (using timing information from both SDMs and RF signals) and the core position were reconstructed with the method described in the previous section. From the total sample of 920 events, 200 were selected in which the position of the core is distributed in a radius of 200m around the center of the station. This was done in order to reduce the computational time of the produced simulations (60 for each event) for applying the method. In figure 5 are shown the distributions

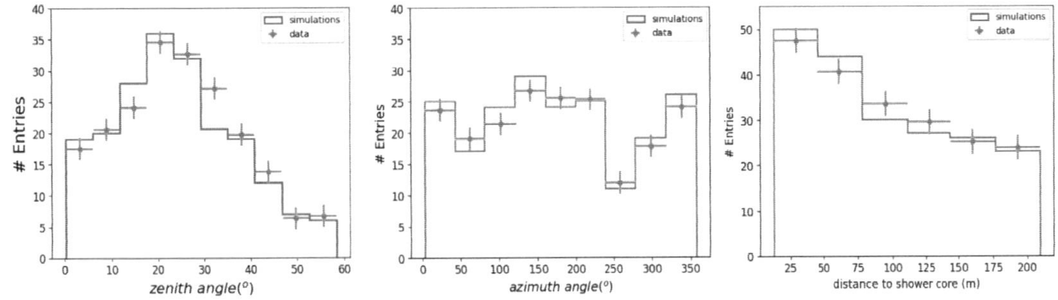

Figure 5. The distributions of the zenith (left), azimuth (center) shower axis angles and the distances from the shower core (right) for the data (red points) and simulations (histogram).

of the zenith (left) and azimuth (center) angles of the reconstructed direction of shower axis. The experimental data are represented with red points, while the simulations with the histogram. In the right plot is depicted the distributions of the distances from the shower core for the data (red points) and simulations (histogram). All data distributions are in good agreement with the corresponding simulations. All the distributions are normalized to the operation time of the station.

5. CGR computation

In this section is described how the CGR can be calculated from the measured fields in the directions EW/NS (E_{ew}, E_{ns}) as well as geometric characteristics of the shower (see figure 6). We analyze the two contributions $\vec{E}_{geo}$ and $\vec{E}_{ch}$ (as projected in the ground plane) in the EW and NS directions

$$E_{ew} = E_{geo}\cos\psi_{geo} + E_{ch}\cos\psi_{ch}, \quad E_{ns} = E_{geo}\sin\psi_{geo} + E_{ch}\sin\psi_{ch} \tag{2}$$

where ψ_{geo} is the angle between the east direction and the direction of the geomagnetic component while the angle ψ_{ch} is the corresponding angle with the direction of the charge excess component (figure 6-right). Both angles are defined to be positive towards north. The $\sin\psi_{geo}$ and $\cos\psi_{geo}$ can be calculated from the projection in the ground plane of the cross product of the unit vector along the shower's propagation direction ($\hat{e}_{v_{sh}}$) and the unit vector in the direction of the geomagnetic field ($\hat{e}_{B_{geo}}$) as given by the equation

$$\begin{aligned} \cos\psi_{geo} &= \left(\hat{e}_{v_{sh}} \times \hat{e}_{B_{geo}}\right)_{ew} = \sin\phi_{sh}\sin\theta_{sh}\cos\theta_{B_{geo}} - \sin\phi_{B_{geo}}\sin\theta_{B_{geo}}\cos\theta_{sh} \\ \sin\psi_{geo} &= \left(\hat{e}_{v_{sh}} \times \hat{e}_{B_{geo}}\right)_{ns} = \cos\theta_{sh}\cos\phi_{B_{geo}}\sin\theta_{B_{geo}} - \cos\phi_{sh}\sin\theta_{sh}\cos\theta_{B_{geo}} \end{aligned} \tag{3}$$

where θ_{sh}, ϕ_{sh} are the zenith and azimuth angles of the EAS axis direction while $\theta_{B_{geo}}$, $\phi_{B_{geo}}$ the corresponding angles of the geomagnetic field direction (in the site of Astroneu $\theta_{B_{geo}} = 34.67^0$ and $\phi_{B_{geo}} = 273.54^0$).

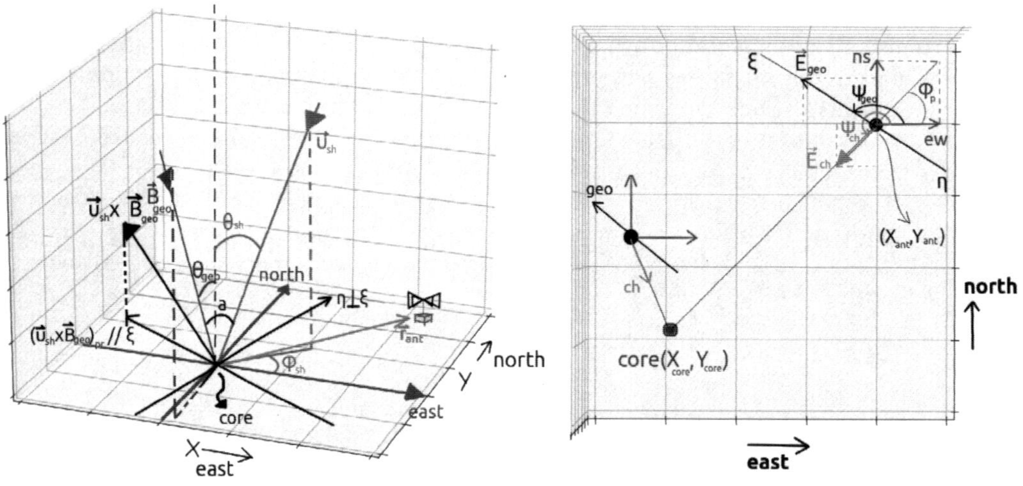

Figure 6. (*Left*) The ground plane geometry of an incoming shower (with zenith and azimuth angles θ_{sh}, ϕ_{sh} respectively) in the direction $\vec{v}_{sh}$ with respect to the direction of the geomagnetic field $\vec{B}_{geo}$. The angle between the shower axis and the geomagnetic field is denoted as α. The projection of the vector $\vec{v}_{sh} \times \vec{B}_{geo}$ in the ground plane is in the direction of the ξ axis. (*Right*) The two electric field components (geomagnetic and charge excess) as projected in the ground plane. The angle between the geomagnetic component ($\vec{E}_{geo}$) with respect to east direction is denoted as ψ_{geo} while the corresponding angle of the charge excess component as ψ_{ch}.

The $\sin\psi_{ch}$ and $\cos\psi_{ch}$ can be derived from the location of the shower core (X_{core}, Y_{core}) in the ground with respect to the location of observation $((X_{ant}, Y_{ant}))$ as shown in figure 6-right

$$\sin\psi_{ch} = \frac{Y_{core} - Y_{ant}}{r_{ant}}, \ \cos\psi_{ch} = \frac{X_{core} - X_{ant}}{r_{ant}}, \ r_{ant} = \sqrt{(X_{ant} - X_{core})^2 + (Y_{ant} - Y_{core})^2} \quad (4)$$

The CGR is defined as the relative strength of the charge excess component to the strength of the geomagnetic component as given by the equation

$$CGR = \frac{|E_{ch}|}{\frac{|E_{geo}|}{|\sin\alpha|}} = \frac{|E_{ch}|}{|E_{geo}|} \cdot |\sin\alpha| \quad (5)$$

The term $\sin\alpha$ reflects the dependence of the geomagnetic component from the angle between the EAS axis and the geomagnetic field and has been added to avoid CGR infinities in case of small geomagnetic angles (α). The final formula used for CGR calculation can be derived combining equations 2, 4

$$\frac{|E_{ns}|}{|E_{ew}|} = \frac{\left|\sin\psi_{geo} + \frac{CGR}{|\sin\alpha|\sin\psi_{ch}}\right|}{\left|\cos\psi_{geo} + \frac{CGR}{|\sin\alpha|\cos\psi_{ch}}\right|} \quad (6)$$

6. CGR data-simulations results

It is predicted from simulations analysis [26] that the CGR increase with increasing opening angle from the emission maximum to the observation location. This implies an increase of CGR with increasing distance from shower axis and decreasing zenith angle. For EAS arriving at larger

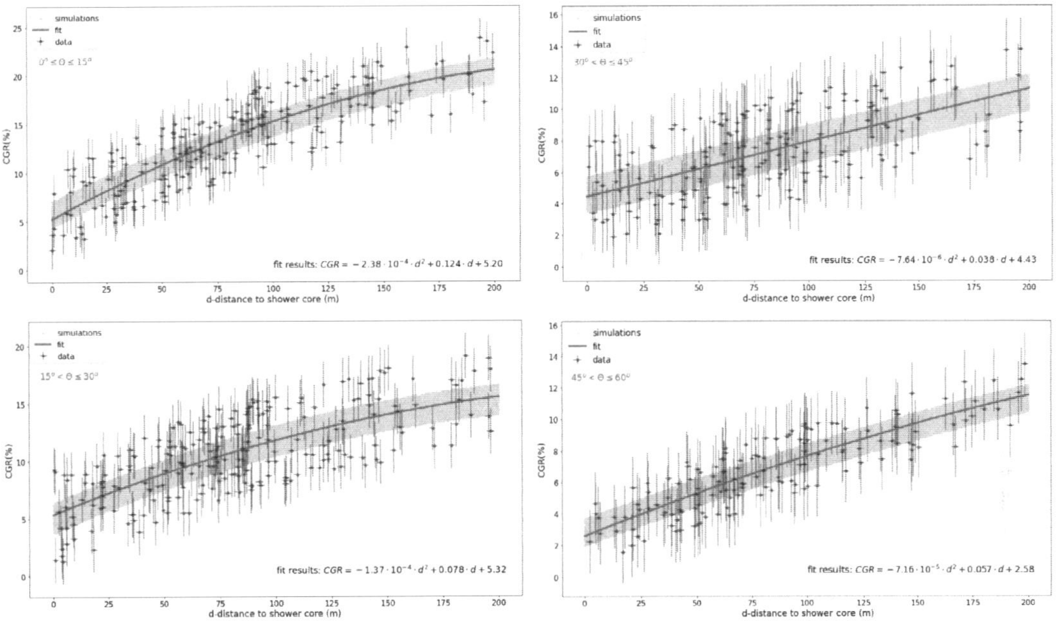

Figure 7. The CGR variation as a function of the distance to shower core for four different zenith angle regions. The blue points correspond to data while the green region to simulations predictions. The red line represents the fit results to data.

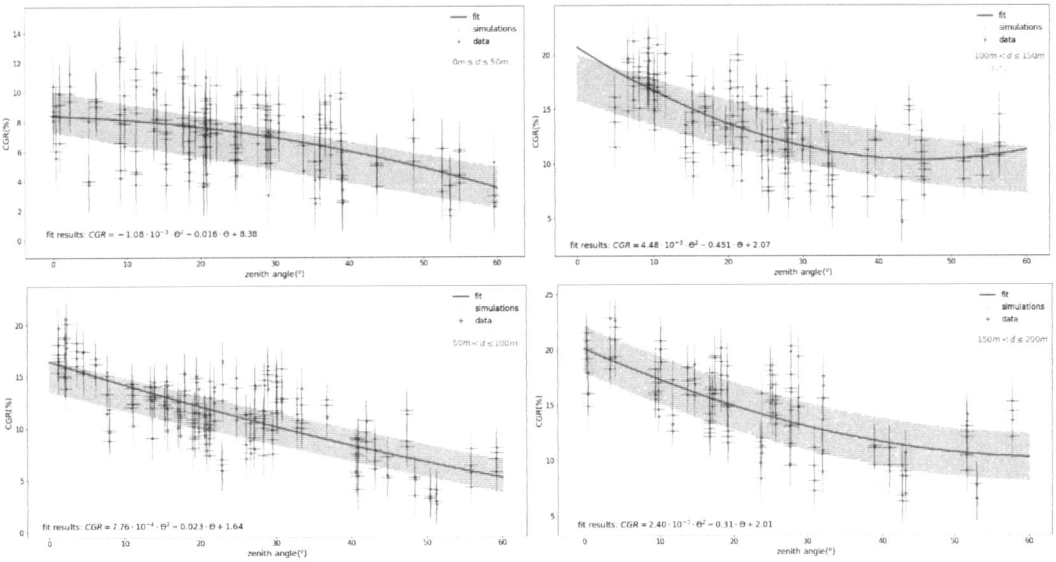

Figure 8. The CGR variation as a function of the zenith angle for four different distance to shower core regions. The blue points correspond to data while the green region to simulations predictions. The red line represents the fit results to data.

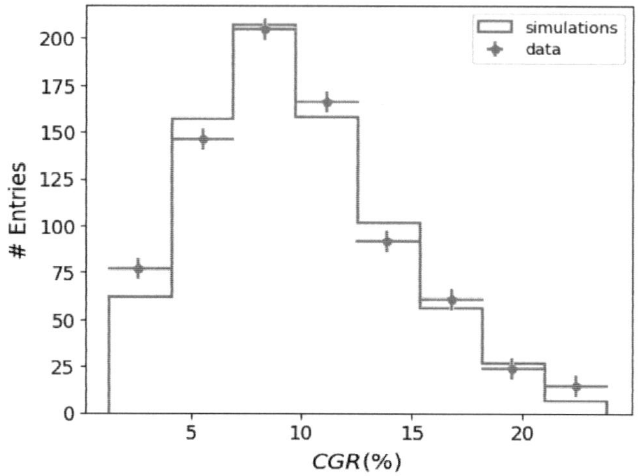

Figure 9. Distribution of the CGR as determined from data (red points) and simulations (histogram). The simulation distribution is normalized to the operation time of the station.

Table 1. The mean values of CGR for different zenith angle and distance to shower core value regions.

	$d \in [0, 50m]$	$d \in (50, 100m]$	$d \in (100, 150m]$	$d \in (150, 200m]$
$\theta_{sh} \in [0, 15^0]$	8.1%	13.15%	17.14%	19.23%
$\theta_{sh} \in (15, 30^0]$	6.96%	10.76%	12.50%	14.92%
$\theta_{sh} \in (30, 45^0]$	5.16%	7.08%	8.74%	10.76%
$\theta_{sh} \in (45, 60^0]$	4.13%	6.56%	8.62%	10.45%

zenith angles the maximum of the RF emission is produced further away from the observer's location and thus the same distance to the shower core corresponds to a smaller opening angle (where the CGR is lower). In order to verify this dependences we calculate the CGR as a function of the distance to the shower core for four different zenith angle bins (from $\theta_{sh} = 0^0$ to $\theta_{sh} = 60^0$ with a step of 15^0). In figure 7 is depicted the corresponding results for both the experimental data (blue points) and simulations (green region) while the red line represents the fit results to data. In all cases the expected increase in the CGR values with increasing distance is clear while data and simulations predictions are in very good agreement.

Furthermore the CGR is calculated as a function of the zenith angle for four different distance to shower core bins (from $d_{core} = 0$ m to $d_{core} = 200$ m with a step of 50 m). In figure 8 is shown the results for both the experimental data (blue points) and simulations (green region) while the red line represents the fit results to data. In all cases the expected decrease in the CGR values with increasing zenith angle is clear while data and simulations predictions are in very good agreement. Finally in figure 9 the distribution of the measured CGR as determined from simulations (histogram) and data (points) are shown. A summary of the mean measured CGR values for different zenith angles and distances is represented in table 1. Since the CGR strongly depends from the EAS core position the obvious coincidence of the CGR values presented in this study compared to previous studies [27, 28] is a strong indication that the reconstruction method is efficient despite the fact that significant improvements can be made (increase the radius around which the core position is searched).

7. Conclusions

Using radio data from the Astroneu's 3SDM-4RF station and a large sample of simulations we have studied a new method for EAS core reconstruction along with charge excess to geomagnetic ratio (CGR) measurements. The measured CGR values range from 3% (inclined shower near the core) to 24% for almost vertical showers at large distances (~ 200 m). The agreement of these results with previous studies is encouraging for the effectiveness of the EAS core determination method. In addition, a study is in progress to improve the efficiency of the method over long distances (> 200 m). In the forthcoming expansion of the Astroneu telescope with more autonomous RF stations CGR measurements are planed be used for an efficient noise rejection algorithm in the self trigger mode. Finally considering that charge excess mechanism is the main contribution for radio emission in dense media, CGR analysis can be used as a tool to distinguish EAS developed in different media.

Acknowledgments

This work was supported by the Hellenic Foundation for Research and Innovation, Project Number 02116.

References

[1] J. V. Jelley et al. 1965 *Nature* **205** 327-328
[2] H. Falcke et al. - LOPES collaboration 2005 *Nature* **435** 313-316
[3] L. Martin et al. Proc.35th ICRC (2017)
[4] P. Schellart et al. 2015 *Astron. Astrophys.* **66** 31
[5] The Pierre Auger Coll. 2016 *Phys. Rev. D* **93** 122
[6] W.D. Apel et al. 2016 *Phys. Lett. B* **763** 179
[7] T. Huege 2016 *Phys. Rept.* **620** 1-52
[8] F. G. Schröder 2017 *Progress in Particle and Nuclear Physics* **93** 1-68
[9] C. Welling et al. 2019 *JCAP* **10** 075
[10] S Nonis et al. 2020 *Phys. Scr.* **95** 084007
[11] F. D. Kahn and I. Lerche 1966 *Royal Society of London Proceedings Series* **A** **289** 206-213
[12] G.A.Askaryan 1962 *Phys. JETP* **14** 441
[13] T. Avgitas et al. 2020 *JINST* **15** T03003
[14] S.E. Tzamarias et al. 2021 *Instrum Exp Tech* **64** 127-140
[15] I. Manthos et al., 2020, *New Astronomy* **81** 101443
[16] A. Leisos et al. 2019 *Universe* **5** 3
[17] S Nonis et al 2021 *J. Phys.: Conf. Ser.* **2105** 012018
[18] S. Nonis et al. 2019 *EPJ Web of Conferences* **210** 05010
[19] D. Charrier, 2012 *Nucl. Instrum. Methods Phys. Res., Sect. A* **662** 142-145
[20] A.G. Tsirigotis et al, 2020 *Eng. Res. Express* **2** 025027
[21] S. Hansen et al. 2004 *IEEE Transactions on Nuclear Science* **51** 926-930
[22] A. Leisos et al, Performance of the autonomous stations of the Astroneu extensive air shower array, arXiv:1801.04768
[23] A. Leisos et al, 2021 *New Astronomy* **82** 101448
[24] V. Marin et al 2012 *Astropart. Phys.* **35** 733
[25] D. Heck et al. 1998 *Forschungszentrum Karlsruhe Report* **FZKA** 6019
[26] K. D. de Vries et al. 2013 *Astroparticle Physics* **45** 23-27
[27] A. Aab et al. 2014 *Phys. Rev. D* **89** 052002
[28] P. Schellart et al. 2014 *JCAP* **10** 014

Journal of Physics: Conference Series **2375** (2022) 012009 doi:10.1088/1742-6596/2375/1/012009

Measurement of the Higgs quartic coupling c_2v from di-Higgs Vector Boson Fusion in the $b\bar{b}\tau^+\tau^-$ channel

Christos Paraskevopoulos

National Technical University of Athens, Heroon Polytechniou 9, Zografou, Greece

E-mail: christos.paraskevopoulos@cern.ch

Abstract. The Brout Englert Higgs (BEH) mechanism of electroweak symmetry breaking and mass generation was experimentally confirmed after the discovery of the Higgs boson at the Large Hadron Collider in 2012. The BEH mechanism not only predicts the existence of a massive scalar particle, but also requires this scalar particle to couple to itself. Double Higgs production provides a unique handle, since it allows the extraction of the trilinear Higgs self-coupling. VBF di-Higgs production also probes the quartic Higgs bosons to vector bosons coupling (c_2v). In this topic the effort on setting constraints on c_2v will be discussed. Event selection and reconstruction will be illustrated as well as a Neural Network designed to identify VBF events.

1. Introduction

The Higgs boson was discovered by the ATLAS and CMS collaborations in 2012 at the Large Hadron Collider [1] [2] . While the Brout-Englert-Higgs(BEH) mechanism predicts the existence of a massive scalar particle, the measurement of the Higgs boson trilinear self-coupling is a crucial validation of the BEH mechanism [3][4] [5][6]. In the Standard Model (SM) pairs of Higgs bosons are mainly produced via gluon-gluon fusion processes via a loop of top quarks and the Higgs self-interaction [7].

In the SM the BEH self coupling λ is uniquely determined by the structure of the scalar potential and is proportional to the product of the heavy-quark Yukawa coupling and the Higgs-boson self-coupling. However the existence of destructive interference between the box and the triangle diagram means di-Higgs production cross section has a small rate of $\sigma_{HH,SM} = 33fb$ at $\sqrt{s} = 13TeV$, about 1000 times smaller than single Higgs production [8]. In addition to constraining λ, in the vector-boson fusion (VBF) channel double Higgs production also probes the strength of the Higgs non-linear interactions with vector bosons at high energies and will allow us to set limits on Higgs quartic coupling c_2v [9]. Moreover, a variety of new physics models predict enhancements to this cross-section [10][11][12]. An example with a new heavy resonance is depicted in Fig.1.

Modifications of the Higgs top-quark Yukawa coupling or modifications of the Higgs self-coupling, or presence of new diagrams with new couplings could enhance the non-resonant di-Higgs production cross section [13] (see Fig.2). The SM HH production is not expected to be observed with the data so far collected by the ATLAS experiment. Beyond Standard Model (BSM) effects could enhance this rate [14]. Therefore, a search for Higgs pair production with current data is a probe of new physics.

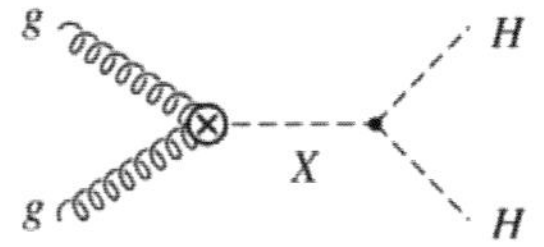

Figure 1. Theories beyond the SM (BSM) predict heavy resonances that could decay into a pair of the SM Higgs bosons, such as a heavy spin-0 scalar X in two-Higgs-doublet models (2HDM).The presence of such resonances would also enhance the di-Higgs production cross section.

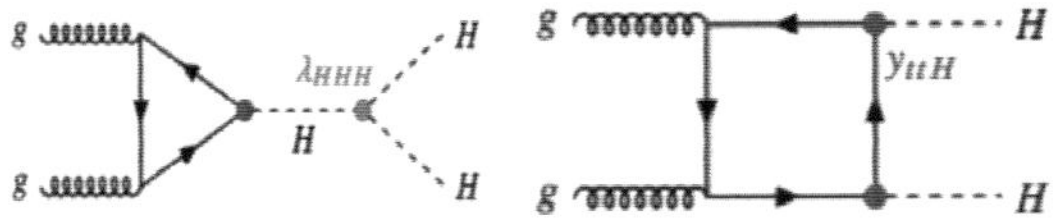

Figure 2. Leading-order Feynman diagrams for the gluon gluon fusion (ggF) non-resonant production of Higgs boson pairs: the "triangle diagram" and the "box diagram". The Higgs boson trilinear self-coupling is denoted λ_{HHH}. The Yukawa coupling of the Higgs to the top-quark is denoted by y_{ttH}

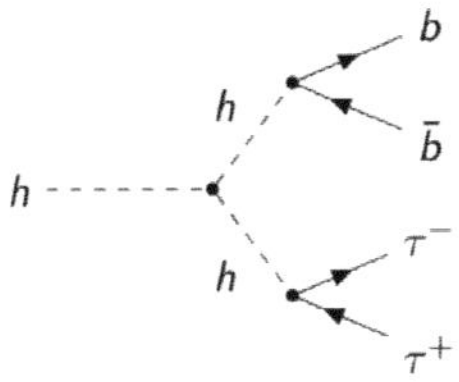

Figure 3. Depiction of the final state of $HH->bb\tau\tau$

The channel we are going to explore is the final state of $b\bar{b}\tau^+\tau^-$, see for example Fig.3. This state is the most sensitive channel with $36fb^{-1}$ data [15]. It represents a relatively clean final state compared to the channels with larger branching fractions ($b\bar{b}b\bar{b}$ and $b\bar{b}W^+W^-$) due to a better separation from the multijet and $\tau\tau$ backgrounds [16][17]. The $b\bar{b}\tau^+\tau^-$ channel has the third largest branching fraction (7.4%) of the experimentally feasible channels [18]. In the final state one of the two τ-leptons of the pair is required to decay hadronically, while the other decays either hadronically ($\tau_{had}\tau_{had}$) or leptonically ($\tau_{lep}\tau_{had}$). While the relatively small background and high branching ratio is an advantage of both decay channels, the *lephad* has a higher trigger efficiency (due to the lepton) while the *hadhad* higher purity (much less $\tau\tau$ background). In this paper we will focus on the *hadhad* channel.

2. What can we access with HH

The main categories we are going to focus are gluon gluon fusion (ggf), see Fig.2 and Vector Boson Fusion (VBF), see Fig.4. The SM predicts non-resonant HH production, with approximately 90% of the total cross-section being due to the ggf process [19].

This channel has been extensively studied in the literature and several final states have been considered, including $b\bar{b}\gamma\gamma$, $b\bar{b}\tau^+\tau^-$, $b\bar{b}W^+W^-$ and $b\bar{b}b\bar{b}$ (for a list of feasibility studies, see for example [20][21][22] [23][24][25]

Working in the infinite top mass approximation, the gluon-fusion di-Higgs production cross

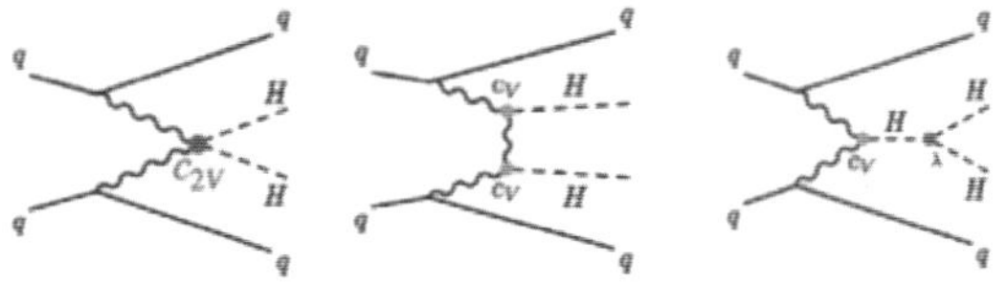

Figure 4. Possible BSM enhancements due to modified coupling strength of c_2v

section was calculated at NLO in [26] and NNLO in [27]. The resummation of soft-gluon emissions was performed at NNLL [28][29]. Recent studies indicate that Higgs pair production in gluon-fusion at the HL-LHC will allow the extraction of the Higgs self-coupling λ with accuracy.

Higgs pairs can also be produced in the VBF channel [9][30][31][32][33] where a soft emission of two vector bosons from the incoming protons is followed by the hard $VV->hh$ scattering, with $V=W,Z$. In the SM, the VBF inclusive cross section at 14 TeV is around 2 fb^{-1} which is more than one order of magnitude smaller than in gluon fusion [9].

Despite its small rate, Higgs pair production via VBF is quite interesting since even small modifications of the SM couplings can induce a striking increase of the cross section as a function of the di-Higgs mass [34].

In the following equation, the parameterization of the effective non-flavor-violating Higgs Lagrangian, in the case the couplings of the Higgs boson to the SM fermions scale with their masses, is given. There is also another parameter, c3, which is not shown:

$$\begin{aligned}\mathscr{L} \supset \frac{1}{2}(\theta_\mu h)^2 - V(h) + \frac{v^2}{4}T_r(D_\mu\Sigma^\dagger D^\mu\Sigma)[1 + 2c_v\frac{h}{v} + c_{2v}\frac{h^2}{v^2} + ...] \\ -m_i\bar{\psi}_{Li}\Sigma(1 + c_\psi\frac{h}{v} + ...)\psi_{Ri} + h.c.,\end{aligned} \quad (1)$$

Assuming that the couplings of the Higgs boson to SM fermions scale with their masses and do not violate flavor, the resulting effective Lagrangian can be parametrized as in equation 1, where $V(h)$ denotes the Higgs potential. The parameters cv, c_2v and $c\psi$ in 1, are in general arbitrary coefficients, normalised so that they equal 1 in the SM. The cv, c_2v and $c3$, which is not shown here, control the strength of the $hVV, hhVV$ and hhh couplings, respectively. A precision model-independent determination of c_2v would provide stringent constraints on a number of BSM scenarios. For generic values of cv and c_2v , the amplitude of the partonic scattering $VV->hh$ grows with the energy $\sqrt{s}$ until the contribution from the new states at the cut off scale Λ unitarizes it. The leading contribution in the energy range $m_W << \sqrt{s} << \Lambda$ comes from the scattering of longitudinal vector bosons and is given by:

$$\Lambda(V_LV_L->hh) \simeq s/v^2(c_2v - c^2v)$$

If the Higgs boson belongs to an electroweak doublet, as suggested by the LHC data, and the modifications to its couplings are small, then the values of c_2v and c^2v are in general predicted to be correlated [35][36] because there is a single dimension-6 effective operator (OH in the basis of [36])which controls the shift in both couplings. Therefore, a high-precision measurement of c_2v can test whether the Higgs boson belongs to a doublet in case a deviation is observed in cv [35].

The main analyses channels are $b\bar{b}\gamma\gamma$, $b\bar{b}\tau^+\tau^-$, and $b\bar{b}b\bar{b}$ where tha last channel full takes advantage of the high bb branching ratio [15][17][37]. Then $b\bar{b}\gamma\gamma$ has excellent trigger and mass resolution for photons. However $b\bar{b}\tau^+\tau^-$ final state is chosen in our case for the following reasons: first, taus are relatively clean, with small background; then, the branching ratio is relatively high and the b-jet tagging is very efficient thanks to the precise reconstruction of

Journal of Physics: Conference Series **2375** (2022) 012009 doi:10.1088/1742-6596/2375/1/012009

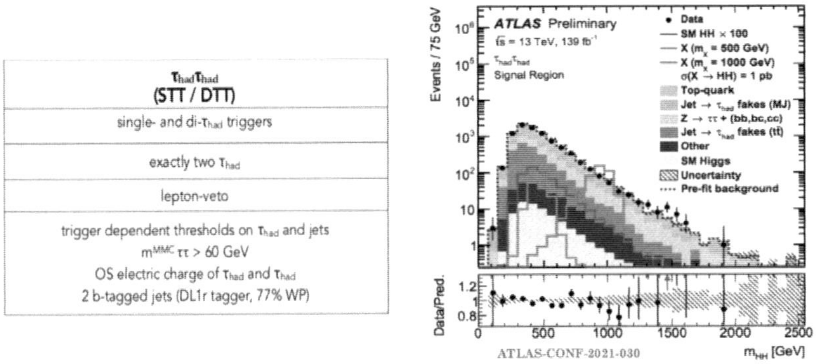

Figure 5. Event selection and backgrounds of $\tau_{had}\tau_{had}$ final state

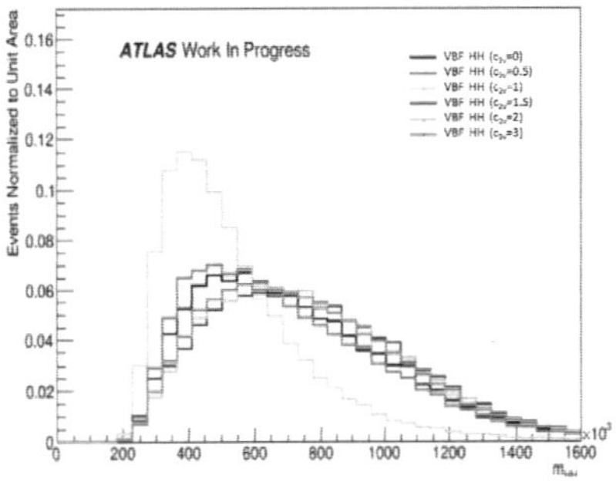

Figure 6. Gaining sensitivity to c_2v is achieved by reconstructing events with large values of m_{HH}

primary and secondary vertices; finally, despite the fact that tau lepton reconstruction might be quite challenging, efficient algorithms have been developed. This enhanced sensitivity to the underlying strength of the Higgs interactions makes double Higgs production via VBF a key process to test the nature of the electroweak symmetry breaking dynamics and to constrain the $hhVV$ quartic coupling. In this work, we revisit the feasibility of VBF Higgs pair production at the LHC and focus on the $HH->b\bar{b}\tau^+\tau^-$ final state.

2.1. VBF jet Event Selection.

The analysis looks for jets with $p_T > 20GeV$ and $|\eta| < 2.5$ which are consistent with the same primary vertex and specifically looks for two tau-jets with the invariance mass of the di-τ system to be $> 60GeV$ and at least one b-jet with $p_T > 45GeV$. The tau-jets are jets with 1 or 3 tracks which are relatively isolated from activity around them, as expected by the fact that taus decay weakly, and b) two b-jets with energies $> 160GeV$, b-jets are jets which are 'b-tagged' because they contain tracks with a significant impact parameter with respect to the beam line, a requirement exploiting the long lifetime of b-quarks. In the VBF analyses we look for two additional jets consistent with the 'forward-backward' VBF topology (see Fig.4). These VBF jets are selected as follows:

- Baseline: Require 2 VBF jets in the event
- Take all forward jets ($|\eta| > 2.5$) with $p_T > 30GeV$.

- Take central jets only ($|\eta| < 2.5$, $p_T > 20GeV$) if they are non-btagged
- Find jet pairing with highest m_{jj} pair and in opposite hemispheres, ($\eta^{j1} * \eta^{j2} < 0$)
- Order VBF jets by p_T

The event selection for the $\tau_{had}\tau_{had}$ final state is shown in Fig.5, together with the event yields expected from the various sources for an integrated luminosity of $139fb^{-1}$ collected by ATLAS at $\sqrt{s} = 13TeV$. The way the di-Higgs mass (m_{HH}) distribution changes as the the c_2v value changes from it SM value ($c_2v = 1$) is shown in Fig.6.

Table 1.

Events	Total	VBF Selection
VBF c2v=0	6.77	4.54
VBF c2v=1	0.167	0.119
VBF c2v=2	4.63	3.125
ggf	5.40	1.57
Comb.bkg	3256.9	

3. Neural Network Design, Hyperparameter Optimization and Results

3.1. The Neural Network design and Input Variables

After comparison of various classification strategies, a Feed Forward Neural Network [38] was chosen in order to classify events in the VBF $bb\tau\tau$ topology. The even-numbered events in the Monte Carlo simulation samples are used to train the Network, which is built with the following characteristics:

- **Classification Strategy**
  - 3-class classification
  - VBF ($c_{2V} = 0$, $c_{2V} = 0.5$, $c_{2V} = 1.5$, $c_{2V} = 2$ - normalized) / ggF / Background
  - Training with even events
- **MultiLayer Perceptron**
  - Input Layer : 9 Neurons (relu)
  - Hidden Layers : $10 * 35$ Neurons (relu)
  - Output Layer : 3 Neuron (softmax)
  - Loss Function : Categorical Crossentropy
  - L2 Regularization for each hidden layer : $3 * 10^{-8}$
  - Loss Optimizer : ADAM
  - Batch size : 16
  - Epochs : 78

3.2. Input variables

- m_{HH} : Di-Higgs invariant mass
- $\Delta R_{\tau\tau}$: The ΔR between the visible-τ decay products
- ΔR_{bb} : The ΔR between the two b-jets

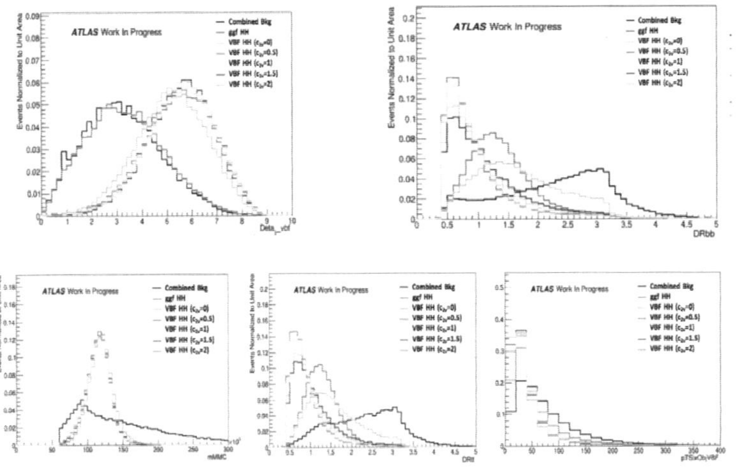

Figure 7. $\Delta\eta_{jj}$, ΔR_{bb}, m_{mmc}, $\Delta R_{\tau\tau}$, $p_T SixObj$ of VBF jets input variables of the NN

- $m_{\tau\tau}^{mmc}$: The invariant mass of the di-τ system, calculated using the MMC momenta of 2 taus [39]
- m_{bb} : The invariant mass of the di-b-jet system
- p_T^{jj} : Di-jet (VBF) transverse momentum
- m_{jj} : The invariant mass of the di-jet(VBF) system
- $\Delta\eta_{jj}$: The $\Delta\eta$ between the two VBF jets
- $\eta_{j1} * \eta_{j2}$: Product of η for two VBF jets
- Δphi_{jj} : The Δphi between the two VBF jets
- $p_T^{\tau\tau}$: The transverse momentum of the two taus
- p_T^{bb} : The transverse momentum of the two bs
- $p_T SixObj$: The transverse momentum of the HH+2-VBF jets

Some of these input variables are shown In Fig7, for background events, ggF HH events, as well as VBF HH events for various combinations of coupling values.

3.3. HyperParameter Optimisation

For the Hyperparameter Optimisation we used a Tree Parzen Estimator with Bayesian approach [40]. The Evaluations of a function are used to predict the next set of good hyperparameters The strategy is summarized in Table.2 and it allows a better overall performance and less time required for optimisation than random search. The Hyperparameters to optimise were:

- Neurons/layer : 20 – 40
- Layers : 6 – 20
- R2 value: $10^9 - 4 * 10^{-7}$
- Batch size: 4 – 24

Different sets were used to perform 50 hyperparameter tests. The optimization variable was the signal for 5 background events.

Table 2. Hyperparameter Optimisation

Hyperparameter	Optimisation strategy	Implementation
Number of hidden layers	Bayesian optimization	TPE
Number of nodes in each layer	Bayesian optimization	TPE
Activation function	Typical suggestion for FF NN	ReLU
Output activation function	Typical suggestion for multiclassification NN	Softmax
Loss function	Typical suggestion for multiclassification NN	Cat. CrossEntropy
Loss minimizer	Dynamic adjustment of per-parameter learning rate	ADAM
Regularisation method	L2 typically works better than L1/Dropout	L2
Regularisation value	Bayesian optimization + fine tuning by hand	TPE
Batch size	Bayesian optimization	TPE
Epochs of training	Best validation loss	EarlyStopping

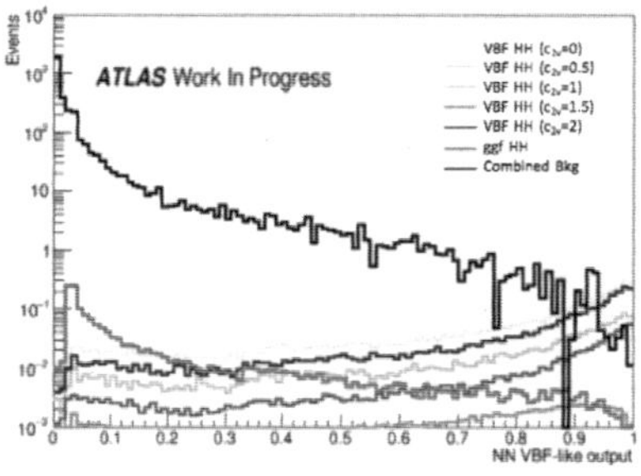

Figure 8. Comparison of various c_2v values of the VBF like ouput, compared with ggf and Combined Background

3.4. Results

The output of the NN described in this section is shown in Fig.8 for Background, ggF HH and VBF HH events, for various values of the c_2v coupling. The number of signal VBF HH, ggF HH and background event expected are shown in Table.1. Imposing a requirement on the minimum NN-output value, changes the remaining event yields and the signal significance is estimated as a function of the minimum NN-output value requirement (see Fig.10). The signal significance is calculated by [41] :

$$Sig = \sqrt{2((s+b)ln(1+s/b)-s)}$$

where s is the signal event yield and b is the sum of the ggF HH and background yields.

Overtraining tests were performed. The two Neural Networks were trained with the optimal Hyperparameters, one trained with the even and the other with the Odd events. Comparison of events is depicted in 10. In order to check that no overtraining is done with this network, two Neural Networks were trained with the optimal Hyperparameters; one trained with the even-numbered and the other with the odd-numbered events and the responses on the even and the odd sample of events were compared (see Fig. 10). Since no overtraining was observed, the results are presented. In order to perform the fit, which will be performed later in our analyses, a the transform tool (https://gitlab.cern.ch/CxAODFramework/TransformTool) will be used. The histograms will be re-binned using an automatic re-binning of the input histograms, which

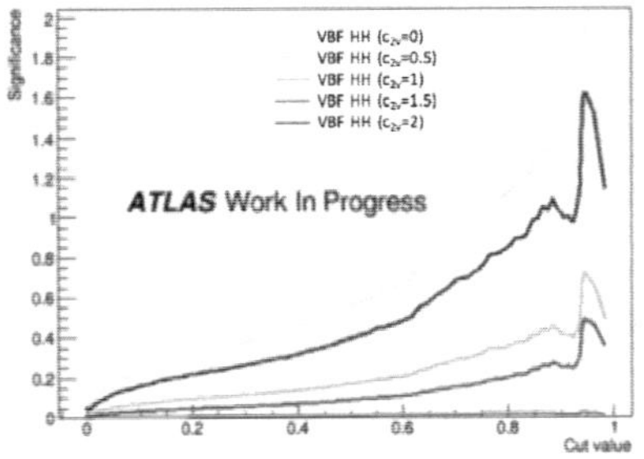

Figure 9. Values of the significance for the different samples Significance is calculated by

merges the histogram bins from right to left until some "pass" criteria are satisfied. In the $\tau_{had}\tau_{had}$ channel of the di-Higgs analysis the histograms are built with 1000 bins. The two chosen functions that will be used are giving better results for the fit performing better not only because they are allowing zero events in the last bin (which is not acceptable as it would cause fit instabilities and would not allow the usage of the asymptotic formula for the limit calculation) so they have been modified adding the requirement of the minimum 5 background events in bin.The fit function chosen as the rebinning function for this channel gives the optimal compromise between good sensitivity and number of bins and events per bin for fit stability. In the case of 5 background events, the NN-output cut which maximizes the signal significance is shown in Table.3, together with the expected signal yield and significance. The expected signal significance is improved considerably, compared to our standard analyses.

Table 3. Significance meausurements

c2v	Significance	Cut value for 5 bkg events	Signal events passing cut	Bkg events passing cut
0	1.082	0.78	2.820 ± 0.022	5.06 ± 0.784
0.5	0.343	0.78	0.842 ± 0.008	5.06 ± 0.784
1	0.019	0.78	0.045 ± 0.004	5.06 ± 0.784
1.5	0.190	0.78	0.0460 ± 0.003	5.06 ± 0.784
2	0.814	0.78	2.076 ± 0.017	5.06 ± 0.784

4. Concusion

This contribution presented studies about accessing the SM Higgs couplings and BSM physics with di-higgs (HH) VBF events in the $bb\tau\tau$ final state. A Feed Forward NN design was chosen to separate the VBF HH signal from ggF HH and background events. A set of good Hyperparameters were found for the NN architecture and additional tests were performed to check if there is overtraining. The signal significance was studied for various c_2v coupling values, with c_2v=1 being the SM value. Since the NN demonstrated a significant improvement in the signal significance, this NN is implemented in our Analysis framework

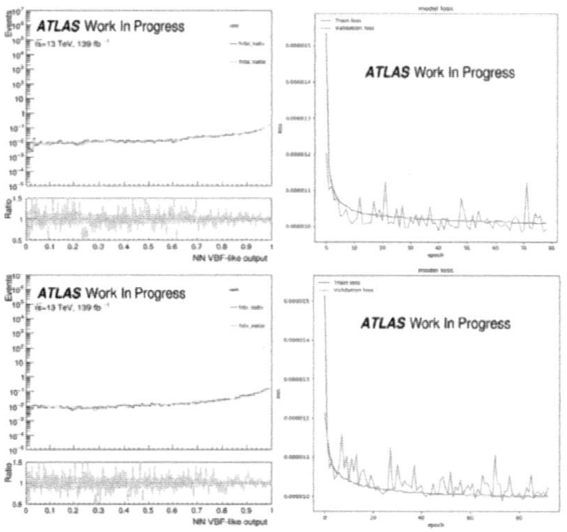

Figure 10. Overtraining tests of the NN. Training was performed with different sets of Even and Odd events

5. Acknowledgments

We acknowledge support of this work by the project "DeTAnet: Detector Development and Technologies for High Energy Physics" (MIS 5029538) which is implemented under the action "Reinforcement of the Research and Innovation Infrastructure" funded by the Operational Programme "Competitiveness, Entrepreneurship and Innovation" (NSRF 2014-2022) and co-financed by Greece and the European Union (European Regional Development Fund).

References

[1] G. Aad, et al., Phys. Lett. B716, 1 (2012). DOI 10.1016/j.physletb.2012.08.020
[2] S. Chatrchyan, et al., Phys. Lett. B716, 30 (2012). DOI 10.1016/j.physletb.2012.08.021
[3] A. Belyaev, M. Drees, O.J.P. Eboli, J.K. Mizukoshi, S.F. Novaes, in Proceedings, International Europhysics Conference on High energy physics (EPS-HEP 1999): Tampere, Finland, July 15-21, 1999 (1999), pp. 748?751. URL http://alice.cern.ch/format/showfull? sysnb=0330978
[4] B. Grinstein, M. Trott, Phys. Rev. D76, 073002 (2007). DOI 10.1103/PhysRevD.76.073002
[5] R. Grober, M. Muhlleitner, JHEP 06, 020 (2011). DOI 10.1007/JHEP06(2011)020
[6] J. Cao, Z. Heng, L. Shang, P. Wan, J.M. Yang, JHEP 04, 134 (2013). DOI 10.1007/JHEP04(2013)134
[7] D. de Florian, et al., Handbook of LHC Higgs Cross Sections: 4. Deciphering the Nature of the Higgs Sector. Tech. rep. (2016)
[8] M. Grazzini et al., Higgs boson pair production at NNLO with top quark mass effects, JHEP 05 (2018) 059, arXiv: 1803.02463 [hep-ph]
[9] R. Contino, C. Grojean, M. Moretti, F. Piccinini, R. Rattazzi, JHEP 05, 089 (2010). DOI 10.1007/JHEP05(2010)089
[10] M. Gouzevitch, A. Oliveira, J. Rojo, R. Rosenfeld, G.P. Salam, V. Sanz, JHEP 07, 148 (2013). DOI 10.1007/JHEP07(2013)148
[11] R. Barbieri, D. Buttazzo, K. Kannike, F. Sala, A. Tesi, Phys. Rev. D87(11), 115018 (2013). DOI 10.1103/PhysRevD.87.115018
[12] U. Ellwanger, JHEP 08, 077 (2013). DOI 10.1007/JHEP08(2013)077
[13] M. Moreno Llacer et al., Study of the CP properties of the top-quark Yukawa interaction in $\tau\bar{\tau}H$ and τH events with $H->\gamma\gamma$, tech. rep. ATL-COM-PHYS-2019-961, CERN, 2019, url: https://cds.cern.ch/record/2684286
[14] G. Branco et al., Theory and phenomenology of two-Higgs-doublet models, Phys. Rept. 516 (2012) 1, arXiv: 1106.0034 [hep-ph]
[15] ATLAS Collaboration, Searches for Higgs boson pair production in the $hh->bb\tau\tau, \gamma\gamma WW, \gamma\gamma bb, bbbb$

channels with the ATLAS detector, Phys. Rev. D 92 (2015) 092004, doi:10.1103/PhysRevD.92.092004, arXiv:1509.04670.
[16] ATLAS Collaboration, Search for Higgs boson pair production in the $bbWW$ decay mode at $\sqrt{s} = 13TeV$ with the ATLAS detector, JHEP 04 (2019) 092, arXiv: 1811.04671 [hep-ex]
[17] ATLAS Collaboration, Search for pair production of Higgs bosons in the $b\bar{b}b\bar{b}$ final state using proton-proton collisions at $\sqrt{s} = 13TeV$ with the ATLAS detector, JHEP 01 (2019) 030, arXiv: 1804.06174 [hep-ex]
[18] A.?M. Sirunyan et al. (CMS Collaboration), Search for Higgs boson pair production in the $bb\tau\tau$ final state in proton-proton collisions at $\sqrt{s} = 8TeV$, Phys. Rev. D 96, 072004
[19] D. de Florian, C. Grojean, F. Maltoni, C. Mariotti, A. Nikitenko, M. Pieri, P. Savard, M. Schumacher, R. Tanaka, Handbook of LHC Higgs cross sections: 4. Deciphering the nature of the Higgs sector, CERN-2017-002-M, DOI: https://doi.org/10.23731/CYRM-2017-002
[20] M.J. Dolan, C. Englert, M. Spannowsky, JHEP 10, 112 (2012). DOI 10.1007/JHEP10(2012)112
[21] A. Papaefstathiou, L.L. Yang, J. Zurita, Phys. Rev. D87(1), 011301 (2013). DOI 10.1103/PhysRevD.87.011301
[22] A.J. Barr, M.J. Dolan, C. Englert, M. Spannowsky, Phys. Lett. B728, 308 (2014). DOI 10.1016/j.physletb.2013.12.011
[23] B. Cooper, N. Konstantinidis, L. Lambourne, D. Wardrope, Phys. Rev. D88(11), 114005 (2013). DOI 10.1103/PhysRevD.88.114005
[24] C.T. Lu, J. Chang, K. Cheung, J.S. Lee, JHEP 08, 133 (2015). DOI 10.1007/JHEP08(2015)133
[25] D. Wardrope, E. Jansen, N. Konstantinidis, B. Cooper, R. Falla, N. Norjoharuddeen, Eur. Phys. J. C75(5), 219 (2015). DOI 10.1140/epjc/ s10052-015-3439-0
[26] S. Dawson, S. Dittmaier, M. Spira, Phys. Rev. D58, 115012 (1998). DOI 10.1103/PhysRevD.58.115012
[27] D. de Florian, J. Mazzitelli, Phys. Rev. Lett. 111, 201801 (2013). DOI 10.1103/PhysRevLett.111.201801
[28] D.Y. Shao, C.S. Li, H.T. Li, J. Wang, JHEP 07, 169 (2013). DOI 10.1007/JHEP07(2013)169
[29] D. de Florian, J. Mazzitelli, JHEP 09, 053 (2015). DOI 10.1007/JHEP09(2015)053
[30] G.F. Giudice, C. Grojean, A. Pomarol, R. Rattazzi, JHEP 06, 045 (2007). DOI 10.1088/1126-6708/2007/06/045
[31] L.S. Ling, R.Y. Zhang, W.G. Ma, L. Guo, W.H. Li, X.Z. Li, Phys. Rev. D89(7), 073001 (2014). DOI 10.1103/PhysRevD.89.073001
[32] M.J. Dolan, C. Englert, N. Greiner, K. Nordstrom, M. Spannowsky, Eur. Phys. J. C75(8), 387 (2015). DOI 10.1140/epjc/s10052-015-3622-3
[33] J. Baglio, A. Djouadi, R. Grber, M.M. Mhlleitner, J. Quevillon, M. Spira, JHEP 04, 151 (2013). DOI 10.1007/JHEP04(2013)151
[34] D.B. Kaplan, H. Georgi, Phys. Lett. B136, 183 (1984). DOI 10.1016/0370-2693(84)91177-8
[35] R. Contino, C. Grojean, D. Pappadopulo, R. Rattazzi, A. Thamm, JHEP 02, 006 (2014). DOI 10.1007/JHEP02(2014)006
[36] G.F. Giudice, C. Grojean, A. Pomarol, R. Rattazzi, JHEP 06, 045 (2007). DOI 10.1088/1126-6708/2007/06/045
[37] ATLAS Collaboration, Search for Higgs boson pair production in the $\gamma\gamma b\bar{b}$ final state with $\sqrt{s} = 13TeV$ collision data collected by the ATLAS experiment, JHEP 11 (2018) 040, arXiv: 1807.04873 [hep-ex]
[38] Daniel Svozil, Vladimr Kvasnicka, Ji Pospichal,Introduction to multi-layer feed-forward neural networks,Chemometrics and Intelligent Laboratory Systems, Volume 39, Issue 1,1997,Pages 43-62,ISSN 0169-7439,https://doi.org/10.1016/S0169-7439(97)00061-0.
[39] A. Elagin, P. Murat and A. Safanov, 'A New Mass Reconstruction Technique for Resonances Decaying to $\tau\tau$', arXiv:1012.4686v2 [hep-ex] 22-Feb-2011
[40] Yoshihiko Ozaki, Yuki Tanigaki, Shuhei Watanabe, Masaki Onishi, Multiobjective tree-structured parzen estimator for computationally expensive optimization problems, GECCO '20: Proceedings of the 2020 Genetic and Evolutionary Computation Conference, June 2020, Pages 533?541, https://doi.org/10.1145/3377930.3389817
[41] Glen Cowan, Kyle Cranmer, Eilam Gross, Ofer Vitells, Asymptotic formulae for likelihood-based tests of new physics, Eur. Phys. J. C (2011) 71: 1554 DOI 10.1140/epjc/s10052-011-1554-0

Journal of Physics: Conference Series **2375** (2022) 012010 doi:10.1088/1742-6596/2375/1/012010

The New Small Wheel High Voltage Infrastructure

I. Drivas-Koulouris *, Efstathios Karentzos**

*National Technical University of Athens, Physics Department, 9 Heroon Polytechniou, Zografos, Postal code 15780, Athens GR

**Albert-Ludwigs-Universität Freiburg, Fahnenbergplatz, 79085 Freiburg im Breisgau, Germany

E-mail: ioannis.drivas-koulouris@cern.ch, efstathios.karentzos@cern.ch

Abstract. The ATLAS Muon Spectrometer is subjected to a series of upgrades in order to face the High Luminosity era of Phase-II. One of the main projects is the New Small Wheel (NSW), which has replaced the existing Small Wheel (SW), and is a perfect complement to the ATLAS Muon Spectrometer (MS) in the Endcap regions. The NSW is the combination of two prototype detector technologies, the small Thin Gap Chambers (sTGC) and the resistive Micromegas (MM). In order to operate smoothly the two technologies, a dedicated hardware infrastructure has been installed and modified accordingly based on the needs of each muon sub-system. In order to control and monitor the hardware for both technologies a Detector Control System (DCS) has been developed, following the architecture of the legacy Muon sub-systems. Aim of this work is the presentation of the development and implementation of these HV DCS systems under the Point 1 schema.

1. Introduction

The replacement of SW with the NSW in the context of the upgrade of MS [1] brought many changes to all the systems involved. The biggest was on DCS [2] where 2 new technologies were introduced in order to meet the prerequisite upgrades. Along with the automatic control system, where it is based on how it is installed the hardware. The MM system has a direct approach by hosting the CAEN A7038 HV boards [3] on a CAEN SY4527 Mainframe [4], while in total 6 mainframes (3 per side) provide the power and service all the HV channels. The sTGC follows the daisy chain logic, where a CAEN SY4527 Mainframe hosts four CAEN A1676A Branch Controller [5]. Each branch controller handles a CAEN EasyCrate3000 and either one channel of a 48V ACDC Generator; CAEN 3485s for service [6] and 3486 for power [7]. The last element of the chain are the HV boards on the crates. All the HV boards have been validated on surface before the installation at the ATLAS cavern. In order this chain of hardware to be operated succesfully, the ATLAS experiment has a dedicated central Detector Control System (DCS), which checks the operational status of different sub-systems and enables the communication between them. Similarly, the NSW follows the same conventions and thus it is being controlled by dedicated projects built with the SIMATIC WinCC Open Architecture (WinCC-OA) [8].

Journal of Physics: Conference Series **2375** (2022) 012010 doi:10.1088/1742-6596/2375/1/012010

2. Finite State Machine (FSM)

The FSM [9] is an abstract machine that is able to be in only one of a finite number of states at a time; the state can change when initiated by a triggering event or condition (transition state). A particular FSM node is defined by a list of states, and the triggering conditions for the transitions between them. Figure 1 depicts the ATLAS FSM architecture. The Global Control Station (GCS), which is in supervision of the overall operation of the detector, provides high level control and monitoring of all the sub-detectors; the execution of the commands and the data processing are managed at the lower levels. Sub-detector Control Stations (SCSs) constitute

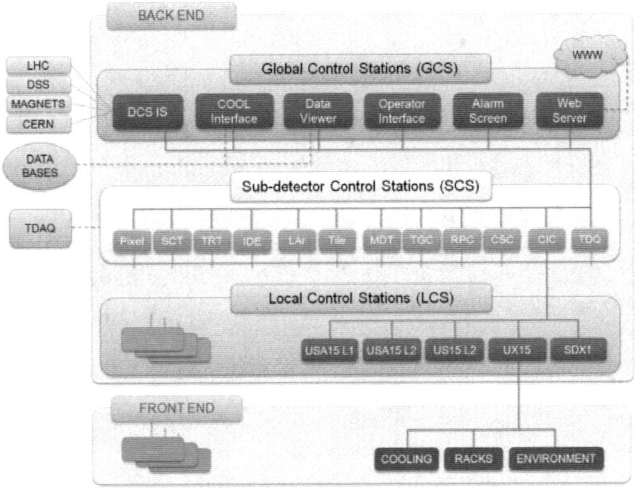

Figure 1. The ATLAS DCS architecture.

the middle level of the "back-end" architecture. The SCS allows the full local operation of a sub-detector.

The DCS activates the supervision of the equipment using commands, processes and actions on the operational parameters of the detector. It is also responsible for the communication between the internal parts of the system and with external control systems. At this level in the hierarchy, the connection with the Data AcQuisition (DAQ) system is established in order to guarantee the synchronization between the detector operation and the physics data taking. For this purpose the sub-detector is divided into partitions which are based on the DAQ TTC (Timing, Trigger and Control) zones. Both systems are synchronized via the DAQ-DCS Communication (DDC) software package. The FSM plays an important role, reporting the DCS state of the TTC zones to the DAQ and executing commands received from DAQ. The final part of the FSM hierarchy on the back-end side, is the Local Control Stations (LCSs). This layer provides the low level control and monitoring of the SCS services, executing all the received commands from it and at the same time is able to trigger autonomously predefined actions if necessary. Each LCS is in charge of a certain system within the sub-detector (i.e. HV, LV, Racks, gas, cooling, etc).

The lowest level is composed of Device Units (DUs) which implement the connection layer between the FSM and the hardware, defining in such way the granularity of the system. Everything below these boundaries is not accesible through the FSM.

3. NSW Caen Easy System

The CAEN EASY system is a modular system of LV and HV power supplies developed by CAEN for the LHC experiments; are grouped into two main categories, the radiation-sensitive parts, and the radiation hard parts. Among its core features are the remote control, the scalability, the compatibility with operation under strong magnetic fields, and radiation harsh environments like the ATLAS cavern (UX15).

The MMG HV system consists of three mainframes per side and the HV boards are installed directly onto them. The main difference between them is that the third mainframe is fully equipped with 16 A7038AP boards serving in such a way all 16 sectors, 4 layers, and the first 5 PCBs (8 in total per layer). The other mainframes follow the same "mapping" scheme, where 4 A7038STP HV boards supply the remaining 3 PCBs. The system is completed by two boards of type A7038STN, which supplies the voltages of the drift panels. The FSM tree consists of three nodes dedicated to each mainframe. The graphical user interface (panel), as illustrated in Figure 2, monitors the vitals (fan speed, status, V1/V0 set flag, etc.) of each mainframe along with the visualisation of the installed board and their main status parameters.

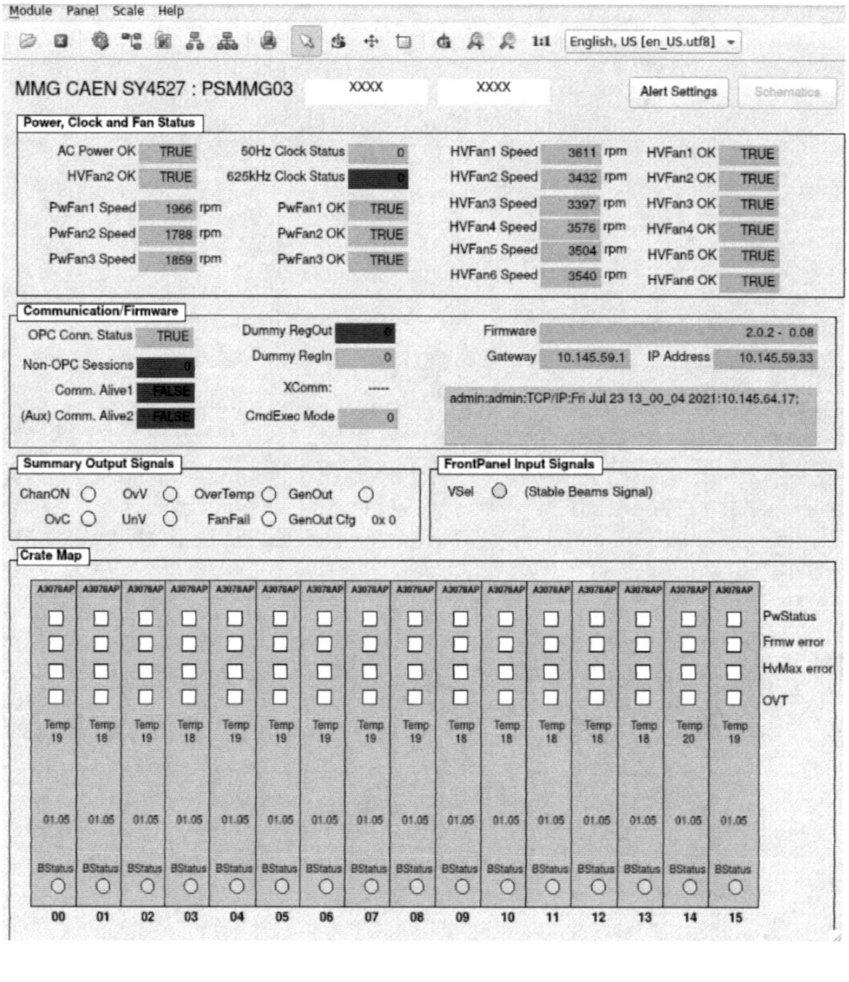

Figure 2. Main FSM MMG panel for the Mainframes

The STG HV system architecture consists of a mainframe, two branch controllers and four crates per side. The branch controllers handles the chain along with the generators. The HV boards are installed in the CAEN EasyCrates and located in the UX15, in a unified format in order to regulate the power consumption per crate. There are two different types of boards, the A3540P and the A3535P serving the 16 STG sectors. There are 4 parts per layer, where the former serves the innermost part of the detector while the other type the rest. The FSM is divided in three nodes :

- one for the mainframe with the main panel similarly to the MMG Tree, but branch controllers' information instead of boards as shown in Figure 3.
- one node branch controllers.
- one node for the crates including also the FSM branch controllers for the HV boards, as shown in Figure 3.

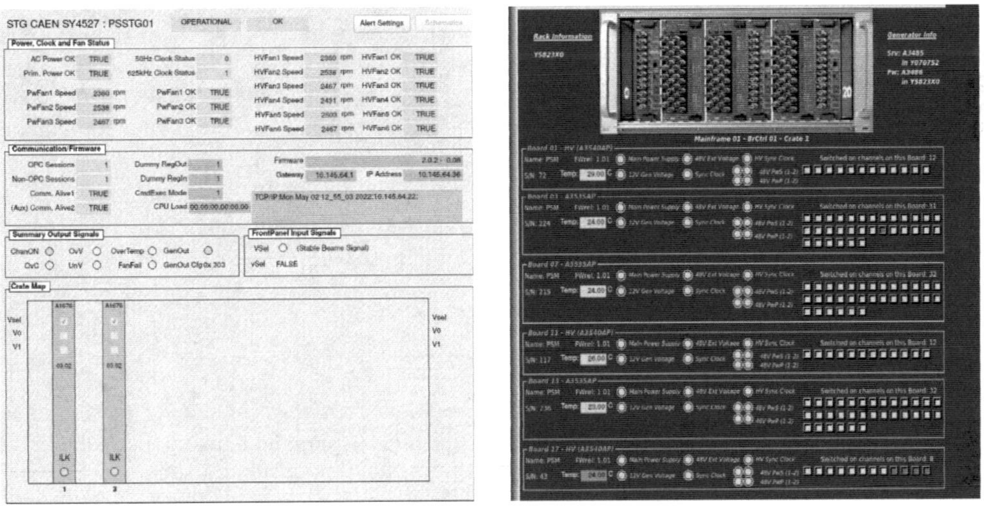

Figure 3. On the left is Main FSM STG panel for the Mainframes. On the right the Main FSM panel for STG crates.

4. NSW (STG) Caen 48V Generators

The 48V generators are used for the STG Sub-detector using the pre-existing muon logic. The crates and boards are supplied with an external DC power of 48V generated by separate AC-DC converters. The system consists of two generators per side of detector. Each generator has two channels. The 48V Service generator is responsible for the control of the easy chain circuitry, distributing the power to the easy crate controllers. The 48V power generator on the other hand is responsible for the output power of the modules up to the channels' level.
Left Figure 4 depicts the associated to the generators rack. Each generator is represented by a rectangle which holds the details about the current, voltage and temperature of every channel. The panel also provides information about the branch controllers and the main logical parameters of the channels as well as supplementary information concerning the connected branch controllers and the generators channels, alongside with a visual representation of the associated to these generators' chambers. The user/expert can select a variety of monitoring controls to perform,

from highlighting the associated to the generator chambers, up to configuring low level elements such as the alert handling. The main STG generators FSM include all these panels, which are associated with the respective nodes. By clicking the state of the nodes, the user can turn on/off the corresponding generators' channels, allowing in such way the control of all or part of the connected hardware.

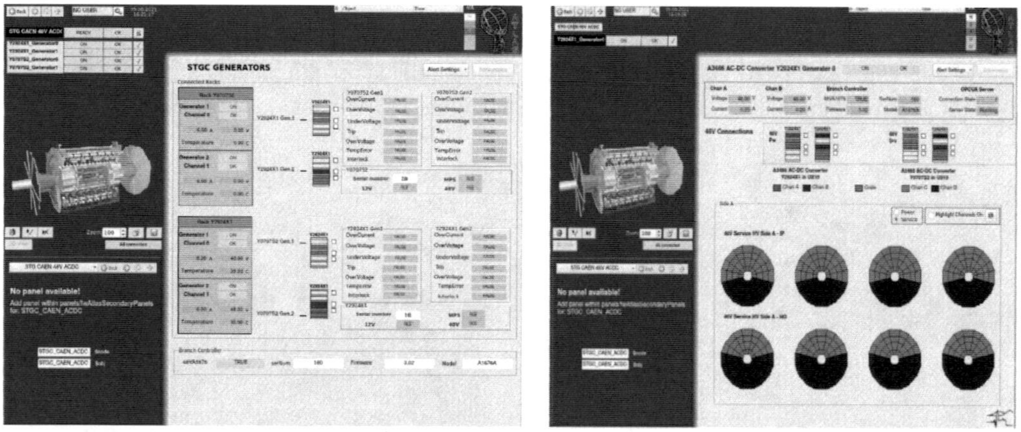

Figure 4. On the left is Main STG generator FSM Panel. On the right Main FSM panel for every individual generator.

5. NSW Reset Network

The CAEN Reset Network was implemented to provide a remote reset during periods without or limited access during beam mode were an actual incident might arise or a clean restart is needed to establish a glitch on power or even communication level. Similarly to the legacy systems, the NSW is using the same logic via the Agilent circuitry [10]. Due to the limitation of the lines in such module, the NSW took some lines from the legacy systems, one from the retired Cathode Strip Chambers (CSC) system, one from the Monitored Drift Tubes (MDT) and one line from the Resistive Plate Chambers (RPC). There the multiplexing of some signals performed in order to include all the hardware of the NSW for both technologies and both board types (LV and HV). A cable is the going from each branch controller to the Agilent module at USA15, where the presence or absence of the 50 Ohm resistor is emulated by a DCS controllable piece of electronics. The Reset Network allows to perform a simultaenous reset of the CAEN mainframes and the branch controllers individually. The MMG FSM has only the mainframe node, shown in Figure 5, where with the "one click" mechanism the expert can reset all the associated to this line mainframes. Unfortunately, due to the limited numbers of lines in the Agilent module, this action will affect all the mainframes of the MMG system for both sides, and thus the expert has to be very careful, communicating to all the affected systems such action in advance.

Similarly to the MMG, the STG reset network has been developed. The possibility of resetting the mainframes of the STG or individually the branch controllers is available. However, the mechanism to independently reset a crate is under development. The FSM of the STG follows explicitly the hardware structure, consisting of two nodes, one for the branch controllers and one for the mainframes as depicted in Figure 6 and 7, respectively. By right clicking the branch controller object element (green circle) the user can perform the following actions:

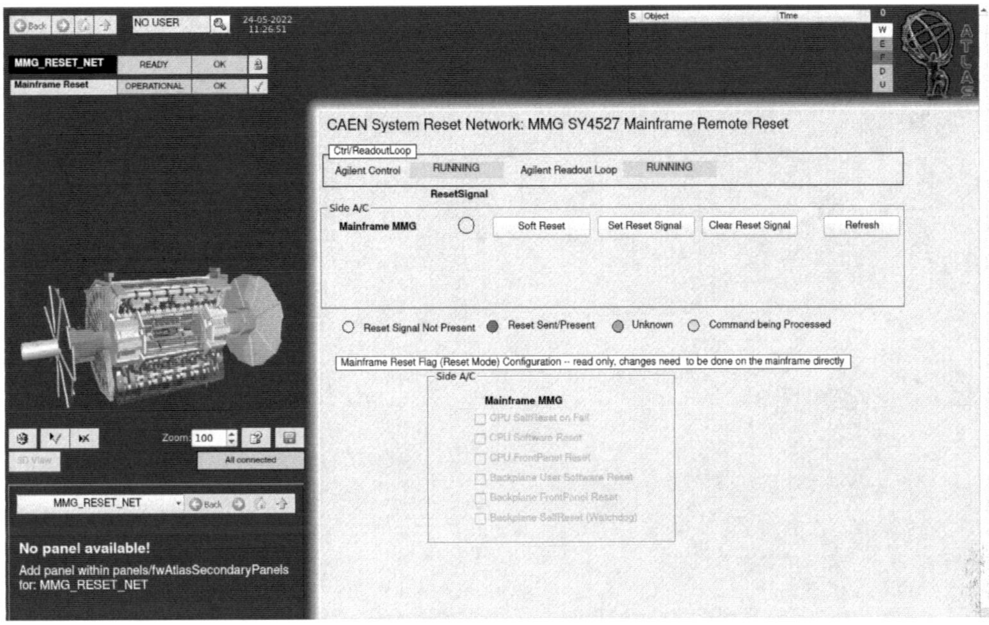

Figure 5. Main panel for MMG Reset Network FSM.

- ENABLE DCS CTRL - Enables reset functionality for this line.
- DISABLE DCS CTRL - Disables reset functionality for this line.
- RESET - Perform the reset for the connected branch controller.
- GET STATUS - Reread the status of the line.
- CANCEL - Abort doing anything on this line.

The STG mainframe Reset panel connects the STG HV with the STG and MMG LV and can be operated by the following actions :

- **Soft Reset**: Reset signal to the mainframe's front-panel reset input is shorter than 1 second, equivalent to shortly depressing the front-panel reset button. A soft reset acts on parts of the mainframe only, as defined in the "Reset Flag" configuration. A soft reset is used to reboot the mainframe CPU without affecting the back plane, i.e. without turning any EASY boards nor any LV/HV channels off.
- **Hard Reset**: This is a full reset, which will turn off all boards and channels controlled of the mainframe. A hard reset occurs if the reset signal is longer than 3 seconds in duration or equivalently, if the reset button is pressed for more than 3 seconds.

The NSW Reset Tree has been tested succesfully validating at the same time the mapping of all the NSW Agilent lines between the mainframes and branches of both HV and LV of the systems.

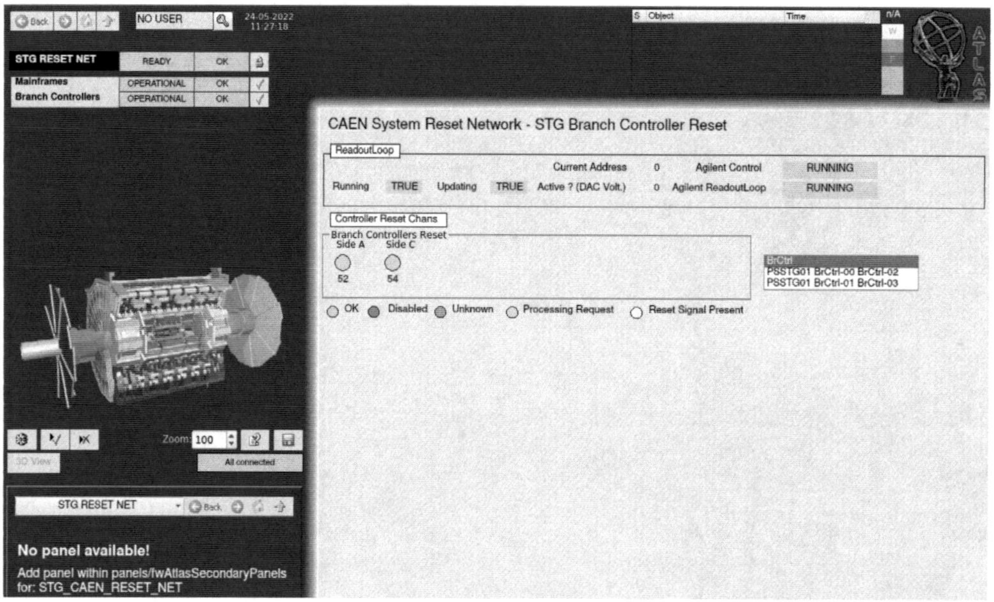

Figure 6. Main panel for STG Reset Network FSM, for controlling the branches.

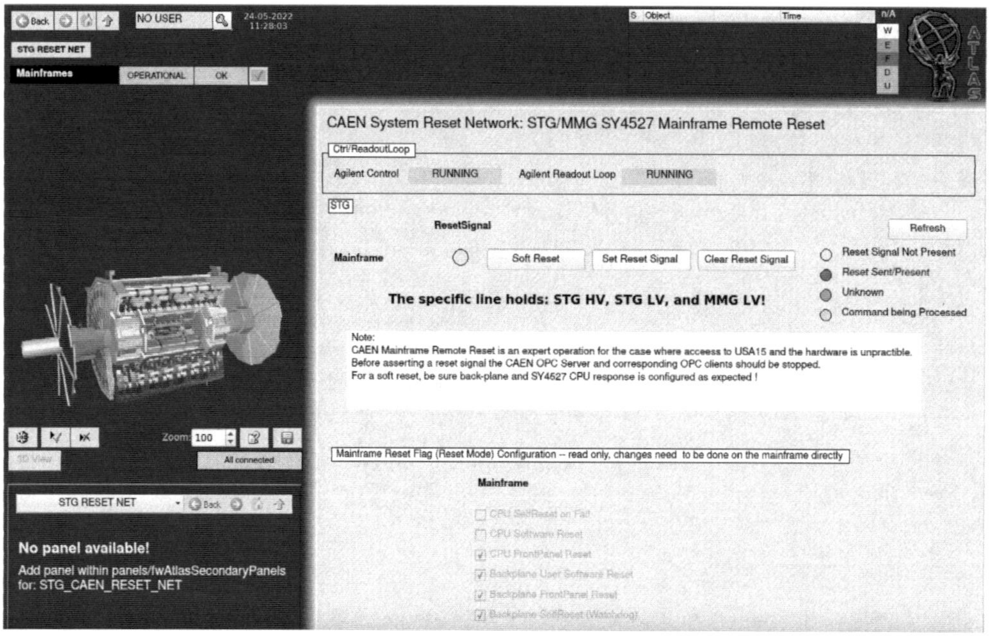

Figure 7. Main panel for STG Reset Network FSM, for controlling the Mainframes

Acknowledgments
We acknowledge support of this work by the project "DeTAnet: Detector Development and Technologies for High Energy Physics" (MIS 5029538) which is implemented under the action "Reinforcement of the Research and Innovation Infrastructure" funded by the Operational Programme "Competitiveness, Entrepreneurship and Innovation" (NSRF 2014–2020) and co-financed by Greece and the European Union (European Regional Development Fund).

References
[1] L. Pontecorvo *The ATLAS Muon Spectrometer* .

[2] A Barriuso Poy *The detector control System of ATLAS* https://iopscience.iop.org/article/10.1088/1748-0221/3/05/P05006/pdfhttps://cds.cern.ch/record/676896/files/sn-atlas-2003-030.pdf.

[3] Ioannis Drivas-Koulouris *Implementation of the DCS System for the validation of MM HV Boards and the DCS System of the new BIS78 Chambers for the upgrade of muon system of the ATLAS Experiment* https://iopscience.iop.org/article/10.1088/1742-6596/2105/1/012027/pdf.

[4] User Manual, *SY4527 - SY4527LC Power Supply Systems* https://www.caen.it/products/sy4527/.

[5] User Manual, *A1676A Branch Controllers* https://www.caen.it/products/a1676a/.

[6] User Manual, *A3485 48V generator* https://www.caen.it/products/a3485/.

[7] User Manual, *A3486 48V generator* https://www.caen.it/products/a3486/.

[8] *SIMATIC WINCC OPEN ARCHITECTURE V3.18* https://assets.new.siemens.com/siemens/assets/api/uuid:2cf67d67-4452-40aa-bbc3-1eef7b469bd8/technical-product-description-wincc-oa-v3-18-en.pdf.

[9] Efstathios Karentzos, *Research and Development of the Micromegas Detector for the New Small Wheel upgrade in the ATLAS Experiment* http://cds.cern.ch/record/2683712/files/CERN-THESIS-2019-094.pdf.

[10] User Manual, *Agilent 34970 Data Acquisition/Switch Unit* https://www.caen.it/products/a3485/.

The Detector Control System of the New Small Wheel Electronics for the ATLAS experiment

Polyneikis Tzanis[1]

[1] National Technical University of Athens, Heroon Polytechniou 9, Zografou, Greece

E-mail: polyneikis.tzanis@cern.ch

Abstract. The present ATLAS Small Wheel Muon detector is be replaced with a New Small Wheel(NSW) detector in order to cope up with the future LHC runs of high luminosit. One crucial part of the integration procedure concerns the validation of the electronics for a system with more than 2.1 M electronic channels. The readout chain is based on optical link technology connecting the back-end to the front-end electronics via the FELIX, which is a newly developed system that serves as the next generation readout driver for ATLAS. For the configuration, calibration and monitoring path the various electronics boards are supplied with the GBT-SCA ASIC and its purpose is to distribute control and monitoring signals to the electronics. Due to its complexity, NSW electronics requires the development of a sophisticated Control System. The use of such a system is necessary to allow the electronics to function consistently, safely and as a seamless interface to all sub-detectors and the technical infrastructure of the experiment. The central system handles the transition between the probe's possible operating states while ensuring continuous monitoring and archiving of the system's operating parameters.

1. ATLAS New Small Wheel

In order to efficiently handle the increased luminosity that will be provided by the High-Luminosity LHC (HL-LHC), the first station of the ATLAS[1] muon end-cap system (Small Wheel, SW) was needed to be replaced. The New Small Wheel (NSW)[2] will have to operate in a high background radiation region (up to $22\,\mathrm{kHz/cm^2}$) while reconstructing muon tracks with high precision as well as providing information for the Level-1 trigger. The detector technologies to be used come from the family of gaseous detectors, the first is called small-strip Thin Gap Chambers (sTGCs), and the second comes from the category of micro-pattern gas detectors and is named Micromesh Gaseous Structure (Micromegas (MM))[3] . The new experimental layout is consisting of 16 detection layers in total and 8 layers per detection technology (8 layers sTGC and 8 layers Micromegas), as shown in Fig. 1. The sTGC detectors are designed to provide fast trigger and high precision muon tracking under the HL-LHC conditions. On the other hand, Micromegas detectors have a small conversion region (5 mm) and fine strip pitch (0.45 mm) resulting in excellent spatial resolution and are primarily used for precise tracking.

2. Electronics Overview

The NSW electronics for the trigger and Data Acquisition (TDAQ) path of both detectors is divided into two major categories, on-detector and off-detector electronics, as shown in Fig. 2. On the left-hand side, the front-end boards that are attached to the chambers bear VMMs, SCAs(Slow Control Adapter) and ROCs(Read Out Controller), while for the sTGCs they also

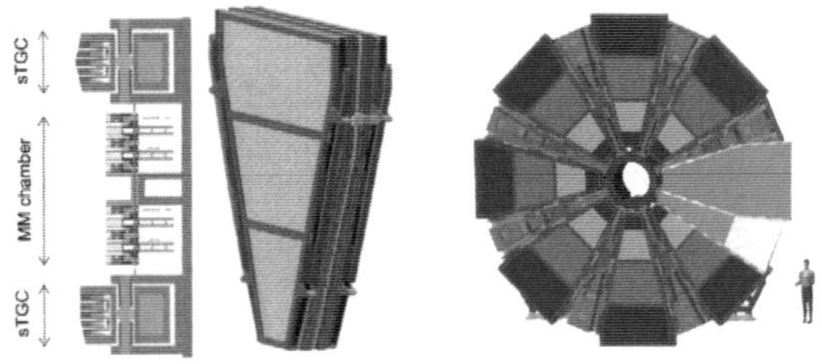

Figure 1. A graphic representation of the NSW sector (left) which consists of 8 layers of Micromegas in the inner part and sandwiched by 4+4 layers of sTGC detectors in the outer parts and view of the NSW (right) with 16 sectors in total.[10]

host TDS chips. The ROC aggregates L1 data from many VMMs[4] and sends them to (Front End LInk eXchange (FELIX)[5] via the GBTx[6]. FELIX also sends trigger signals to the front-end electronics via the ATLAS TTC system. It also sends tracking data from the ROC to the swROD (here depicted to send the data fragments to the HLT), and communicates with the Detector Control System (DCS)[7] for slow control purposes. The Micromegas trigger data are collected from many VMMs by the ART ASIC, which sends them towards the Micromegas trigger processor, alongside geographical and timestamp information. The TDS chips on the other hand, handle the sTGC trigger primitives, and alongside the Pad Trigger and Router boards, are part of the more complex sTGC trigger chain, which has the associated trigger processor FPGA on the back-end as a final destination of its data-flow. Both FPGAs create muon candidates that are then transmitted to the SL. The NSW on-detector electronics (Front-End boards (MMFE8), Level-1 Data Driver Card (L1DDC), ART (Address in Real Time) Data Driver Card (ADDC)) are placed inside the cavern (detector area with radiation and magnetic fields) and consists of custom-made boards mainly using radiation-tolerant Application Specific Integrated Circuits (ASICs). The communication between these boards are established with the use of mini Serial Attached Small Computer System Interface (SCSI) cables. The off-detector electronics FELIX, trigger processor, sector logic and services running on commercial server computers like Read Out Drivers (ROD), Detector Control System (DCS), event monitoring, configuration, trigger monitor and calibration) are placed outside the cavern in an area that is called USA15.

3. FELIX

The keystone of the ATLAS DAQ system will be the FELIX which is an FPGA-based system housed by a commercial server. FELIX will essentially be a bridge between the front-end electronics of all ATLAS detector subsystems, and their corresponding back-end components, which is mostly be software-based, as shown in Fig. 3. Situated in the USA15, FELIX connects to the front-end electronics of the ATLAS cavern via optical links, or GBT links, each one of which is running at 4.8 Gb/s. For the NSW case, FELIX interfaces with the front-end nodes over 24 optical links. These links carry the GBT frame, which is 84-bit wide. The Giga-Bit Transceiver (GBT) protocol is a transmission scheme that involves radiation-tolerant ASICs that are capable of handling the large amounts of data of high energy physics experiments.

4. GBT-SCA

The GBT-SCA ASIC (Giga-Bit Transceiver - Slow Control Adapter)[8] is an integrated circuit built in a commercial 130 nm CMOS technology and is the part of the GBT chipset which purpose is to distribute control and monitoring signals to the front-end electronics embedded in

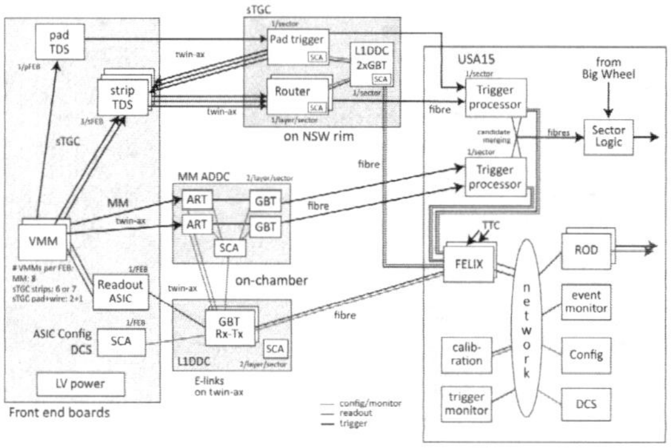

Figure 2. Overview of the NSW electronics scheme. The front-end detector boards are depicted on the left (for MM and sTGC), the data-driver cards (L1DDC, ADDC) in the middle while the back-end electronics can be seen on the right.[10]

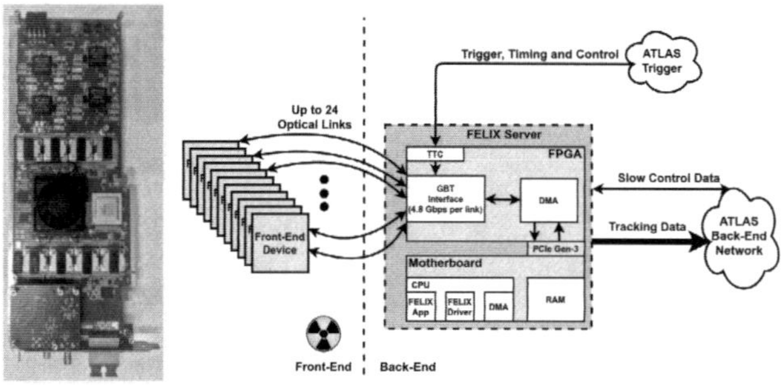

Figure 3. Left: The FELIX BNL712 FPGA board, which the NSW uses for its needs. It features a Xilinx® Kintex Ultrascale XCKU115 FPGA, a PCIe connector to interface with the CPU and the back-end network, and an optical coupler supporting up to 24 links. A dedicated mezzanine board receives the reference clock and the trigger information from the ATLAS TTC system. Right: FELIX and its relationship with other parts of the DAQ system.

the detectors. It connects to a dedicated electrical port on the GBTx ASICs through an 80 Mbps dual redundant bidirectional data-link; namely the e-links. In order to meet the requirements of different front-end ASIC in various experiments, the SCA provides a number of user-configurable electrical interface ports, able to perform concurrent data transfer operations. The user interface ports are: 1 SPI master, 16 independent I2C masters, 1 JTAG master and 32 general-purpose IO signals with individual programmable direction and interrupt generation functionality. It also includes 31 analog inputs multiplexed to a 12 bit ADC featuring offset calibration and gain correction as well as four analog output ports controlled by four independent 8-bits DACs. An illustration of the different blocks are displayed on Fig. 4.

5. SCA OPC-UA Server

For the NSW project alone, $\approx$ 7000 SCAs are used to configure and monitor various ASICs that are part of the general DAQ scheme. All of these SCAs communicate with the back-end (i.e. FELIX) via the GBTx, which implements the GBT protocol in its logic, while FELIX implements the GBT-FPGA core in order to communicate with the aforementioned package.

Journal of Physics: Conference Series **2375** (2022) 012011 doi:10.1088/1742-6596/2375/1/012011

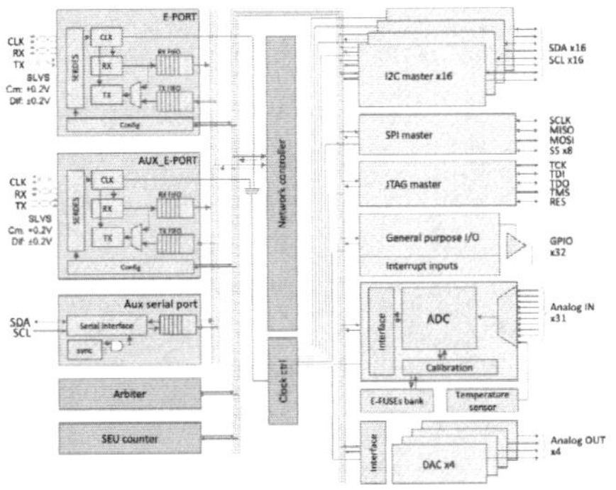

Figure 4. Overview of the GBT-SCA block diagram.[8]

The SCA-to-GBTx interface is the E-link, a serial differential line running at 80 Mb/s over HDLC encoding. The protocol defined by the SCA's requirements is request/respond; that is, for every packet received by the SCA, the back-end that originally transmitted the packet awaits for an associated reply by the SCA. In this SCA-GBTx-FELIX communication chain, the last two components can be viewed as data mediators, so there is one piece missing: the back-end logic that actually builds the packets-to-be-transmitted to the SCA, and handles the inbound traffic from the ASIC. This is a software suite[9], which is a dedicated Open Communications Platform Unified Automation (OPC-UA) server. Overview of the on/off-detector electronics and the DCS, Configuration and Calibration path via the SCA, the FELIX and the SCA OPC UA Server is illustrated in Fig. 5. A common readout path and a separate trigger path are developed for each detector technology. The communication chain starts in the control room with an OPC-UA client, which is the first step into our detector control system. Configuration and monitoring data is requested and sent from here to the OPC-UA server and next to the FELIX PCs, which lie in the underground service area. Then, through radiation-hard fibers, we reach the GBTx chips, which implement the optical links, to finally arrive to the SCA.

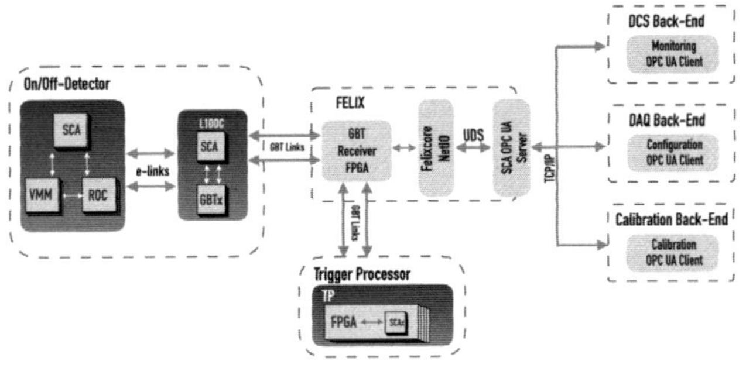

Figure 5. Representation of the on/off-detector electronics and the DCS, Configuration and Calibration path via the SCA, the FELIX and the SCA OPC UA Server [10]

6. ATLAS Detector Control System

The ATLAS DCS has the task to permit coherent and safe operation of ATLAS and to serve as a homogeneous interface to all sub-detectors and the technical infrastructure of the experiment. The DCS must bring the detector into any desired operational state, continuously monitor and archive the operational parameters, signal any abnormal behavior. The DCS was designed and implemented within the frame of the Joint Controls Project (JCOP), a collaboration of the CERN controls group and DCS teams of the LHC experiments. Standards for DCS hardware and software were established together with implementation guidelines both, commonly for [11] and specifically for ATLAS. It combines common standards for the use of DCS hardware based on SCADA system Siemens, WinCC Open Architecture, also known as PVSS with its older name, where it serves as the basis for all DCS applications. Fig. 6 depicts the architecture of DCS which can be divided into Front-End (FE) equipment and a system Back-End (BE) [12]. FE includes equipment DCS, including customized electronic systems and related services such as high voltage power supplies and cooling circuits. The BE system uses the software WinCC, integrating front-end control systems into the framework components to facilitate the integration of standard hardware devices and the implementation of homogeneous control applications. The two ends of DCS communicate mainly through the industrial protocol bus CAN, while the communication standard OPC is used as a communication software protocol. Communication between subsystems of the system is handled by WinCC via a local area network. A distributed finite state machine, Finite State Machine (FSM), represents the complete hierarchy of the BE system as it integrates more than 10 million data elements into a single tree structure and ensures proper operation and efficient handling errors in each operating layer. The most important element of this system is the structure datapoint (DP), which plays the role of the global variable network. Each element of this structure has a unique name and configurability, and special DPs are used to read data from the hardware components updated by the interface OPC client-server communication.

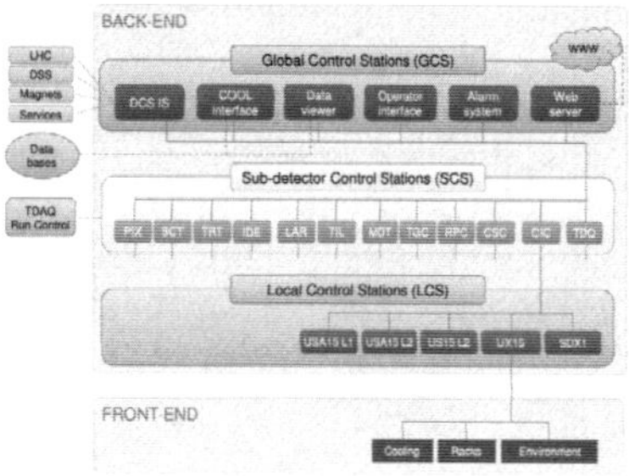

Figure 6. Graphical representation of the DCS architecture of the experiment ATLAS divided into 3 levels: GCS, SCS and LCS.[7]

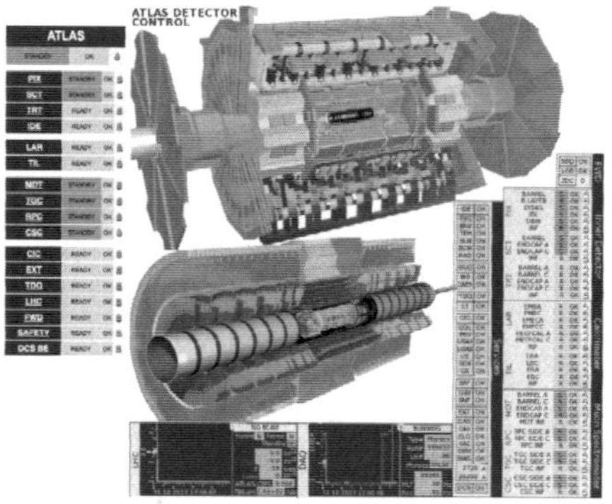

Figure 7. Operator interface (FSM Screen) showing the detector in STANDBY configuration during LHC ramp-up. The top hierarchy level object is shown together with its children objects (top left) and the associated main panel (bottom right). The UI allows the navigation to any FSM object and associated panel within the whole ATLAS DCS hierarchy. On the top right, a list of objects with non-OK Status allow shortcut navigation to problems.[7]

7. NSW Detector Control System

The scope of the NSW DCS design is to be integrated inside the ATLAS DCS and provide continues monitoring and robust interface for all the NSW hardware. Since MMG&STG are muon detector, they will be part of the Muon DCS system which consists of the following sub-detectors:

- Monitored Drift Tubes (MDT)
- Resistive Plate Chambers (RPC)
- Thin Gap Chambers (TGC)
- Cathode Strip Chamber (CSC)

Due to the NSW Upgrade project, the CSC will be removed and the corresponding DCS as a sequence. For the NSW purpose, two 2 sub-detector will be added under the MUO FSM tree:

- MMG for Micromegas
- STG for sTGC

Since the NSW is a big project, the various hardware must be divided into categories for a smooth architecture design. Thus, the NSW consists of the following hardware blocks:

- High Voltage (HV): consists of the hardware which provide the high voltage to the detector which is important for the muon detector and operation of the detector;
- Low Voltage (LV): consists of the hardware which which provide the low voltage to the electronics;
- MDT-DCS-Module (MDM): consists of the hardware interface which hosts the temperature and bField sensors;
- Electronics (ELTX): are the electronics which are responsible for the data acquisition, trigger and the data transfer;
- Gas: consists of the hardware for the gas of the detector which is important for the normal operation of the detectors;
- Cooling: consists of the hardware which provides the appropriate cooling of the electronics;
- Virtual Machine Environment (VME): is the hardware which hosts the modular electronics;
- Advanced Telecommunications Computing Architecture (ATCA): is the hardware which hosts the modular electronics;
- Detector safety system (DSS): is the hardware interface which is important for the detector safety;
- Beam Interlock System (BIS): is the interface which provide the beam interlock flags which are important for the physics run and LHC operation.

An overview of the integrated MMG&STG DCS within the ATLAS and Muon DCS is displayed in Fig. 8.

8. Electronics Control Station

8.1. System setup

The electronics system setup is consisting of 28 FELIX servers, 12 for MMG and 16 for STG accordingly. The FELIX-SCA OPC UA Server granularity is be 1-to-1 meaning that we have 1 SCA OPC UA Server per sub-detector sector. An example is illustrated on Fig. 9. For the system operation, automated sequence has to be settled up on the FELIX machine in order to perform the GBT e-link configuration, the FELIX API initialization and the SCA OPC UA Server automatically in the FELIX machine boot.

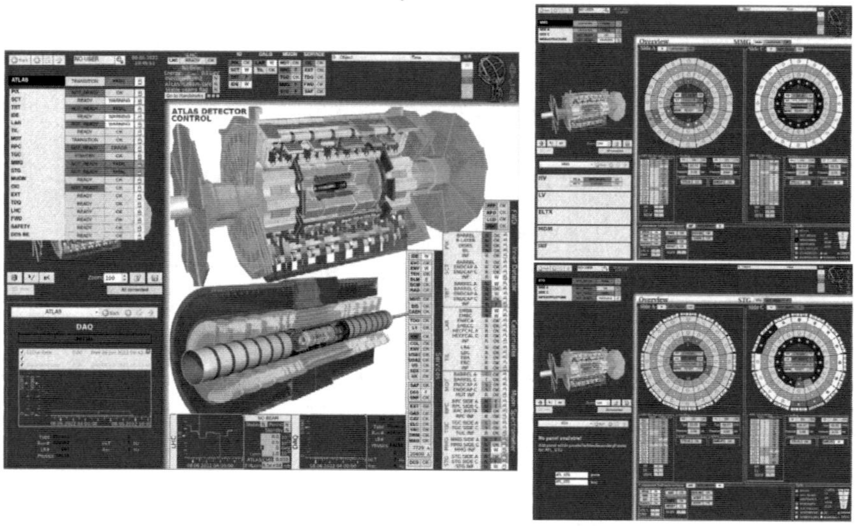

Figure 8. An overview of the integrated MMG& STG DCS within the ATLAS and Muon DCS.

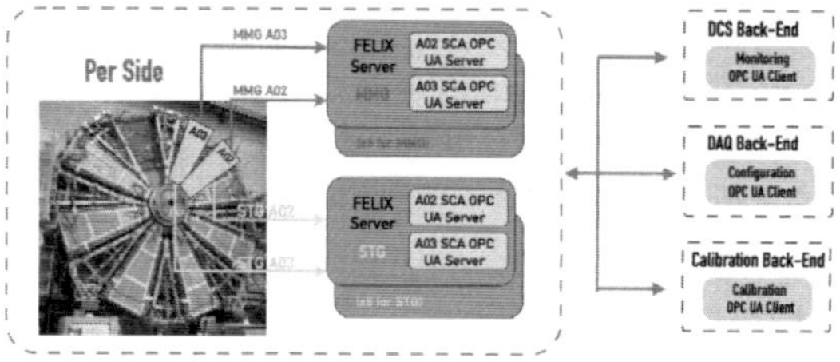

Figure 9. An illustration of the FELIX-SCA OPC UA Server granularity mapping.

8.2. Configuration

The NSW is a fully autonomous trigger and tracking detector system, adequately supported by an advanced electronics scheme and ready to handle the challenges of increased instantaneous luminosity at the High Luminosity LHC. It includes more than 60k front-end ASICs and a few tens of FPGAs, which need to be configured before every run of the experiment. This process needs to be efficient and quick, since it needs to happen a few times per day. The main ASIC to be configured is the VMM, the front-end signal pre-processing chip that can readout 64 channels. A few kbytes of configuration data include thresholds for each channel, but also global registers to define the gain, time-to-amplitude conversion and many others. For this procedure we need to use the SCA's SPI master to communicate with the 8 SCA's SPI slaves in the VMMs. Also, the SCA's GPIO interface is required, to act as an enable signal. Moreover, a similar scheme using the SCA's I2C interface is used for configuration of the TDS chip, which is used for timing and trigger.

8.3. Calibration

The calibration procedure consists of various timing and charge calibrations of the front-end electronics. Gain calibration is done by varying the signal input using the internal pulser of the chip. A specific configuration file needs to be loaded on the VMM chips, which is done with the SCA. Calibration of a time-to-amplitude converter is done by skewing the input clock, which is performed by re-configuring the specific on-board with different settings. Baseline and noise level is defined by reading out each channel output with the ADC when no collisions are occurring.

8.4. Monitoring

Due to its complexity and long-term operation, the ATLAS detector requires the development of an advanced DCS for the electronics monitoring using the SCA chip, which is installed on the 7000 front-end boards of the NSW. The use of such a system is necessary for the safe operation of the detector as well as to act as a homogeneous interface to all the sub-detectors and the technical infrastructure of the experiment. This system gives us the ability to monitor more than 100000 parameters which include all the power/temperature sensors, on-chip temperature and information, which are connected to the SCA on all the front-tend boards of the NSW. In order to achieve the electronics monitoring via the FELIX-SCA and the SCA OPC UA server chain some steps must be performed in advance. The first step is the creation of the SCA OPC UA Server XML file, which the various sensors connected to the SCA ADC channels are specified. Then, the second step is the initialization of the FELIXcore and the SCA OPC UA Server, services which are running into the FELIX machine. The third step is the subscription to the SCA OPC UA Server via a WinCC-OA OPC UA Client. Then, another step is the creation of the SCA OPC UA Server XML items into DPEs inside the WinCC-OA. And finally, the monitor of the board parameters is accessible via the SCA electronics control station . The electronics control station has been developed, following the existing look, feel and command architecture of the other Muon sub-systems, in order to facilitate the shifter/expert operations. It is mapped onto a hierarchy of Finite State Machine (FSM) elements using the toolkit. For each individual layer, a main panel has been developed, providing the user with useful information, reflecting the state and status of the detector, its vitals displayed in trend-plots; while a secondary panel provides supplementary details. A general view of the graphical user interface of the top FSM nodes and their constituents is shown in the Fig. 10.

Figure 10. Left: The interface of the Electronics control station and in particular the layer view of the sector which user can navigate and see in a glance the electronics status. Right: The sector view of the Electronics control station.

Each board's monitoring and conditions can be displayed on the panel. The state and status of the board depends on the status data point element of each board's temperature and power sensors and the communication validity of the SCA OPC UA server. The projects were validated through daily usage from shifters in the commissioning site at a first stage and then through the NSW control in ATLAS Control room.

8.5. DCS-DAQ interaction

As already mentioned above, DCS, calibration and configuration share the common SCA OPC UA Server path for the SCA. Although, the front-end boards which are equipped with the VMM, are using the same SCA ADC input channel for both monitoring and calibration purposes. During the Physics run, the VMM monitor output is equivalent to the VMM temperature but during the Calibration run, the VMM monitor output displays the baseline measurement which is used for the board noice calibration measurements. Thus, during calibration run, VMM shows fake temperature values so a interface plugin between DCS-DAQ should be implemented in order to prevent the DCS to trigger several alarms and warnings. The solution found via the common SCA OPC UA Server and the FreeVariable status which is mainly a user defined OPC-UA item which can be controlled and monitored both by DCS and DAQ back-ends applications. Thus, during the Calibration run, the DCS monitors the configuration status for this specific VMM of the FEB and the alarm will be disabled corresponding FEB's VMM, as shown in Fig. 11. In addition, more DCS-DAQ interaction tools are on-going in order to monitor the FELIX status, the GBTx alignment information and various FELIX infrastructure parameters.

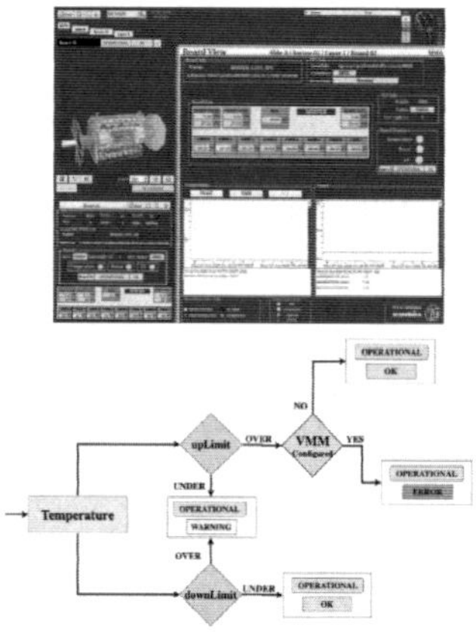

Figure 11. Top: Overview of the board panel of the electronics control station which shows the VMM temperature alarm disabled using the configuration status of the FEB's VMM via the SCA OPC UA Server. Bottom: A flow chart of the logic for the status and state definition based on the VMM temperature and the configuration status.

Acknowledgments
We acknowledge support of this work by the project "DeTAnet: Detector Development and Technologies for High Energy Physics" (MIS 5029538) which is implemented under the action "Reinforcement of the Research and Innovation Infrastructure" funded by the Operational Programme "Competitiveness, Entrepreneurship and Innovation" (NSRF 2014–2020) and co-financed by Greece and the European Union (European Regional Development Fund).

References
[1] ATLAS Collaboration, *The ATLAS Experiment at the CERN Large Hadron Collider*, JINST 3 (2008) S08003.

[2] ATLAS Collaboration, *New Small Wheel Technical Design Report*, CERN-LHCC-2013-006, ATLAS-TDR-020, `https://cds.cern.ch/record/1552862`

[3] T. Alexopoulos et al , *Performance studies of resistive-strip bulk micromegas detectors in view of the ATLAS New Small Wheel upgrade*, Nucl. Instrum. Meth. A 937 (2019) 125.

[4] G. Iakovidis, V. Polychronakos, G. de Geronimo, *VMM - An ASIC for Micropattern Detectors*, ATL-MUON-PROC-2015-015, `https://cds.cern.ch/record/2104297`

[5] W Wu *FELIX: the New Detector Interface for the ATLAS Experiment* IEEE Transactions on Nuclear Science vol. 66 no. 7 pp 986-992 July 2019 doi: 10.1109/TNS.2019.2913617.

[6] P. Moreira, A. Marchioro, K. Kloukinas, *The GBT : a Proposed Architecure for Multi-Gb/s Data Transmission in High Energy Physics*, CERN-2007-007, `https://cds.cern.ch/record/1091474` JINST 7 (2012) C12022.

[7] Barriuso A P 2008 et al. *The detector control system of the atlas experiment* JINST 3 P05006

[8] A. Caratell et al. , *The GBT-SCA, a radiation tolerant ASIC for detector control and monitoring applications in HEP experiments*, JINST 10 (2015) C03034.

[9] P. Moschovakos et al. , *A Software Suite for the Radiation Tolerant Giga-bit Transceiver - Slow Control Adapter*, ICALEPCS Conf.Proc. WEPHA102.2020

[10] P. Tzanis, *Electronics performance of the ATLAS New Small Wheel Micromegas wedges at CERN*, JINST 15(2020) C07002.

[11] Holme O et al. 2005 *The JCOP framework* ICALEPCS Conf.Proc.C051010:WE2.1-6O

[12] P. Tzanis, *The Detector Control System of the New Small Wheel for the ATLAS experiment*, IOP Publishing 2105 012025

Journal of Physics: Conference Series **2375** (2022) 012012 doi:10.1088/1742-6596/2375/1/012012

Formulating E- & T-Model Inflation in Supergravity

C Pallis

Laboratory of Physics, Faculty of Engineering,
Aristotle University of Thessaloniki, Thessaloniki
GR-541 24, Greece

E-mail: kpallis@gen.auth.gr

Abstract. We present novel realizations of E- and T-model inflation within Supergravity which are largely associated with the existence of a pole of order one and two respectively in the kinetic term of the inflaton superfield. This pole arises due to the selected logarithmic Kahler potentials K, which parameterize hyperbolic manifolds with scalar curvature related to the coefficient $(-N) < 0$ of a logarithmic term. The associated superpotential W exhibits the same R charge with the inflaton-accompanying superfield and includes all the allowed terms. The role of the inflaton can be played by a gauge singlet or non-singlet superfield. Models with one logarithmic term in K for the inflaton, require $N = 2$, some tuning – of the order of 10^{-5} – between the terms of W and predict a tensor-to-scalar ratio r at the level of 0.001. The tuning can be totally eluded for more structured K's, with N values increasing with r and spectral index close or even equal to its present central observational value.

1. Introduction

It is well-known [1–3] that the presence of a pole in the kinetic term of the inflaton gives rise to inflationary models collectively named α-attractors [4, 5]. Confining ourselves to poles of order one and two, the lagrangian of the homogenous inflaton field $\phi = \phi(t)$ can be written as

$$\mathcal{L} = \sqrt{-\mathfrak{g}}\left(\frac{N_p}{2f_p^2}\dot\phi^2 - V_{\rm I}(\phi)\right) \quad \text{with} \quad f_p = 1-\phi^p,\ p=1,2 \ \text{ and } \ N_p > 0. \tag{1}$$

Also, $\mathfrak{g}$ is the determinant of the background Friedmann-Robertson-Walker metric $g^{\mu\nu}$ with signature $(+,-,-,-)$, dot stands for derivation *with respect to* (w.r.t) the cosmic time and $V_{\rm I} = V_{\rm I}(\phi)$ is a polynomial function of ϕ. Expressing the canonically normalized field, $\widehat\phi$, in terms of ϕ as follows

$$\frac{d\widehat\phi}{d\phi} = J = \frac{\sqrt{N_p}}{f_p} \quad \Rightarrow \quad \phi = \begin{cases} 1 - e^{-\widehat\phi/\sqrt{N_1}} & \text{for } p=1, \\ \tanh\left(\frac{\widehat\phi}{\sqrt{N_2}}\right) & \text{for } p=2\,, \end{cases} \tag{2}$$

we easily infer that $V_{\rm I}(\widehat\phi)$ develops a plateau for $\widehat\phi \gg 1$, and so it becomes suitable for driving inflation of chaotic type. It is called *E-Model Inflation* (EMI) [5, 6] (or α-Starobinsky model [7]) and *T-Model Inflation* (TMI) [6, 8] for $p = 1$ and 2 respectively.

In this talk, based on Ref. [2, 3], we present novel realizations of *E and/or T-Model Inflation* (ETI) in the context of *Supergravity* (SUGRA). Namely, in Sec. 2 we describe our general strategy in establishing those models and then – in Sec. 3 – we specify three models (δEM, EM2, EM4) employing a gauge singlet inflaton and three (δTM, TM4, TM8) using a gauge non-singlet inflaton. We end up with our numerical results in Sec. 4 and our conclusions in Sec. 5.

Throughout the text, the subscript $,\chi$ denotes derivation w.r.t the field χ, charge conjugation is denoted by a star (*) and we use units where the reduced Planck scale $m_{\rm P} = 2.44 \cdot 10^{18}$ GeV is set equal to unity.

2. Strategy of the SUGRA Embedding

We start our investigation presenting the basic formulation of scalar theory within SUGRA in Sec. 2.1 and then – in Sec. 2.2 and 2.3 – we outline our methodology in constructing viable ETI.

2.1. General Formalism

The part of the SUGRA lagrangian including the (complex) scalar fields z^α can be written as

$$\mathcal{L} = \sqrt{-\mathfrak{g}}\left(K_{\alpha\bar\beta}D_\mu z^\alpha D^\mu z^{*\bar\beta} - V_{\rm SUGRA}\right), \tag{3}$$

where the kinetic mixing is controlled by the Kähler potential K and the relevant metric defined as

$$K_{\alpha\bar\beta} = K_{,z^\alpha z^{*\bar\beta}} > 0 \;\;\text{with}\;\; K^{\bar\beta\alpha}K_{\alpha\bar\gamma} = \delta^{\bar\beta}_{\bar\gamma}. \tag{4a}$$

Also, the covariant derivatives for the scalar fields z^α are given by

$$D_\mu z^\alpha = \partial_\mu z^\alpha + igA^{\rm a}_\mu T^{\rm a}_{\alpha\beta}z^\beta \tag{4b}$$

with $A^{\rm a}_\mu$ being the vector gauge fields, g the (unified) gauge coupling constant and $T^{\rm a}$ with ${\rm a} = 1, ..., {\rm dim}G_{\rm GUT}$ the generators of a *Grand Unified Theory* (GUT) gauge group $G_{\rm GUT}$. Here and henceforth, the scalar components of the various superfields are denoted by the same superfield symbol.

The SUGRA scalar potential, $V_{\rm SUGRA}$, is given in terms of K, and the superpotential, W, by

$$V_{\rm SUGRA} = V_{\rm F} + V_{\rm D} \;\;\text{with}\;\; V_{\rm F} = e^K\left(K^{\alpha\bar\beta}D_\alpha W D^*_{\bar\beta}W^* - 3|W|^2\right) \;\;\text{and}\;\; V_{\rm D} = g^2{\textstyle\sum_{\rm a}}{\rm D}^2_{\rm a}/2. \tag{5}$$

The Kähler covariant derivative and the D-terms read

$$D_\alpha W = W_{,z^\alpha} + K_{,z^\alpha}W \;\;\text{and}\;\; {\rm D}_{\rm a} = z_\alpha\left(T_{\rm a}\right)^\alpha_\beta K^\beta \;\;\text{with}\;\; K^\alpha = K_{,z^\alpha}\,. \tag{6}$$

Therefore, ETI introduced in Sec. 1 can be supersymmetrized, if we select conveniently the functions K and W so that Eq. (3) reproduces Eq. (1).

2.2. Inflaton's Sector

We concentrate on ETI driven by $V_{\rm F}$ assuring that $V_{\rm D} = 0$ during it. This condition may be attained identifying the inflaton either with (the radial part of) a gauge singlet superfield $z^2 := \Phi$ or with the radial part of a conjugate pair of Higgs superfields, $z^2 := \Phi$ and $z^3 := \bar\Phi$.

To achieve a kinetic term in Eq. (3) similar to that in Eq. (1) for $p = 1$ and 2, we need to establish suitable K's so that (during ETI)

$$\langle K\rangle_{\rm I} = -N\ln f_p \quad\text{and}\quad \langle K_{\alpha\bar\beta}\rangle_{\rm I} = N/f_p^2 \tag{7}$$

with N related to N_p. However, from the F-term contribution to Eq. (5), we remark that K affects – besides the kinetic mixing – $V_{\rm SUGRA}$, which, in turn, depends on W too. Therefore, f_p is generically expected to emerge also in the denominator of $V_{\rm SUGRA}$ making difficult the establishment of an inflationary era. This problem can be addressed [2, 3] either by tuning the terms of W so that the pole is removed from $V_{\rm SUGRA}$ thanks to cancellations or adopting a more structured K which yields the desired kinetic terms in Eq. (1) but lets $V_{\rm SUGRA}$ immune.

2.3. Stabilizer's Sector
We reserved $\alpha = 1$ for a gauge singlet superfield, $z^1 = S$ called stabilizer or goldstino, which assists [10] us to formulate ETI of chaotic type in SUGRA. We assume that S and W are equally charged under a global R symmetry and so it appears linearly in W multiplying its other terms. Also, it generates for $\langle S\rangle_{\rm I} = 0$ the inflationary potential via the only term of $V_{\rm SUGRA}$ which remains alive

$$V_{\rm I} = \langle V_{\rm F}\rangle_{\rm I} = \langle e^K K^{SS^*}|W_{,S}|^2\rangle_{\rm I}, \tag{8}$$

where the symbol "$\langle Q\rangle_{\rm I}$" denotes the value of a quantity Q during ETI. Such an adjustment assures the boundedness of $V_{\rm I}$, whereas S can be stabilized at $\langle S\rangle_{\rm I} = 0$, if we select [11]

$$K_2 = N_S \ln\left(1 + |S|^2/N_S\right) \;\;\Rightarrow\;\; \langle K_2^{SS^*}\rangle_{\rm I} = 1 \;\text{ with } 0 < N_S < 6. \tag{9}$$

K_2 parameterizes the compact manifold $SU(2)/U(1)$. Note that for $\langle S\rangle_{\rm I} = 0$, S is canonically normalized and so we do not mention it again henceforth.

3. Inflationary Settings
We here explain the construction of the EMI and TMI in Sec. 3.1 and 3.2 respectively.

3.1. E-Model Inflation
This setting is realized in presence of two gauge singlet superfields S and Φ. The relevant SUGRA setup is presented in Sec. 3.1.1 whereas the resulting models are derived in Sec. 3.1.2.

3.1.1. SUGRA Set-up We adopt the most general renormalizable W consistent with the R symmetry mentioned in Sec. 2.3, i.e.,

$$W = S\left(\lambda_1\Phi + \lambda_2\Phi^2 - M^2\right) \tag{10}$$

where λ_1, λ_2 and M are free parameters. As regards K, this includes, besides K_2 in Eq. (9), one of the following K's:

$$K_{1\rm s} = -N\ln\left(1 - \Phi/2 - \Phi^*/2\right) \quad\text{or}\quad \widetilde{K}_{1\rm s} = -N\ln\frac{(1 - \Phi/2 - \Phi^*/2)}{(1-\Phi)^{1/2}(1-\Phi^*)^{1/2}}, \tag{11}$$

with $\mathsf{Re}(\Phi) < 1$ and $N > 0$. For the considered versions of EMI, δEM and EM2– EM4, the total K respectively is

$$\text{(a) } K_{21\rm s} = K_2 + K_{1\rm s} \;\text{ and }\; \text{(b) } \widetilde{K}_{21\rm s} = K_2 + \widetilde{K}_{1\rm s}. \tag{12}$$

The elimination of pole in $V_{\rm I}$ for δEM can be achieved if we set

$$N = 2 \quad\text{and}\quad r_{21} = -\lambda_2/\lambda_1 \simeq 1 + \delta_{21} \quad\text{with}\quad \delta_{21} \sim 0 \quad\text{and}\quad M \ll 1. \tag{13}$$

No denominator exists in $V_{\rm I}$ for EM2 and EM4 and so N, λ_1, λ_2 and M are free parameters.

3.1.2. Inflationary Configuration The inflationary potential can be derived from Eq. (8) specifying the inflationary trajectory as follows

$$\langle S\rangle_{\rm I} = 0 \;\text{ and }\; \langle\theta\rangle_{\rm I} := \mathsf{arg}\langle\Phi\rangle_{\rm I} = 0. \tag{14}$$

Inserting the quantities above into Eq. (8) and taking into account Eq. (9) and

$$\langle e^K\rangle_{\rm I} = \begin{cases} f_1^{-N} & \text{for } K = K_{21\rm s}, \\ 1 & \text{for } K = \widetilde{K}_{21\rm s}, \end{cases} \tag{15}$$

Fields	Eigen-states	Masses Squared	
			$K = K_{21s}$ $K = \widetilde{K}_{21s}$
1 real scalar	$\widehat{\theta}$	$\widehat{m}^2_\theta$	$6H_{\rm I}^2$
2 real scalars	$\widehat{s}_1,\ \widehat{s}_2$	$\widehat{m}^2_s$	$6H_{\rm I}^2/N_S$
2 Weyl spinors	$(\widehat{\psi}_\Phi \pm \widehat{\psi}_S)/\sqrt{2}$	$\widehat{m}^2_{\psi\pm}$	$6n(1-\phi)^2H_{\rm I}^2/N\phi^2$

Table 1. *Mass spectrum of our EMI along the trajectory of Eq. (14) – we take $n = 1$ for δEM and EM2 whereas $n = 2$ for EM4.*

we arrive at the following master equation

$$V_{\rm I} = \lambda^2 \begin{cases} \left(\phi - r_{21}\phi^2 - M_1^2\right)^2/f_1^N & \text{for } \delta\mathsf{EM}, \\ \left(\phi - r_{21}\phi^2 - M_1^2\right)^2 & \text{for } \mathsf{EM2}, \\ \left(\phi^2 - r_{12}\phi - M_2^2\right)^2 & \text{for } \mathsf{EM4}, \end{cases} \tag{16}$$

where $\phi = \mathsf{Re}(\Phi)$, $r_{ij} = -\lambda_i/\lambda_j$ with $i, j = 1, 2$ and λ and M_i are identified as follows

$$\lambda = \begin{cases} \lambda_1 \ \text{ and } \ M_1 = M/\sqrt{\lambda_1} & \text{for } \delta\mathsf{EM} \text{ and } \mathsf{EM2}, \\ \lambda_2 \ \text{ and } \ M_2 = M/\sqrt{\lambda_2} & \text{for } \mathsf{EM4}. \end{cases} \tag{17}$$

As advertised in Sec. 3.1.1, the pole in f_1 is presumably present in $V_{\rm I}$ for δEM, but it disappears for EM2 and EM4. The arrangement of Eq. (13), though, renders the pole harmless for δEM.

We introduce the canonically normalized fields, $\widehat{\phi}$ and $\widehat{\theta}$, as follows

$$\langle K_{\Phi\Phi^*}\rangle_{\rm I}|\dot{\Phi}|^2 \simeq \frac{1}{2}\left(\dot{\widehat{\phi}}^2 + \dot{\widehat{\theta}}^2\right) \ \Rightarrow\ \frac{d\widehat{\phi}}{d\phi} = J = \frac{\sqrt{N/2}}{f_1} \ \text{ and } \ \widehat{\theta} \simeq J\phi\theta \ \text{ with } \ \langle K_{\Phi\Phi^*}\rangle_{\rm I} = \frac{N}{4f_1^2}. \tag{18}$$

We see that the relation between ϕ and $\widehat{\phi}$ is identical with Eq. (2) for $p = 1$, if we do the replacement $N_1 = N/2$.

To check the stability of $V_{\rm SUGRA}$ in Eq. (5) along the trajectory in Eq. (14) w.r.t the fluctuations of z^α's, we construct the mass spectrum of the theory. Our results are summarized in Table 1. Taking the limit $\delta_{21} = M_1 = 0$ for δEM, $r_{21} = M_1 = 0$ for EM2 and $r_{12} = M_2 = 0$ for EM4, we find the expressions of the masses squared $\widehat{m}^2_{\chi^\alpha}$ (with $\chi^\alpha = \theta$ and s) arranged in Table 1. We there display the masses $\widehat{m}^2_{\psi\pm}$ of the corresponding fermions too – we define $\widehat{\psi}_\Phi = J\psi_\Phi$ where ψ_Φ and ψ_S are the Weyl spinors associated with S and Φ respectively. We notice that the relevant expressions can take a unified form for all models – recall that we use $N = 2$ in δEM – and approach, close to $\phi = \phi_\star \simeq 1$, rather well the quite lengthy, exact ones employed in our numerical computation. From them we can appreciate the role of $N_S < 6$ in retaining positive $\widehat{m}^2_s$. Also, we confirm that $\widehat{m}^2_{\chi^\alpha} \gg H_{\rm I}^2 \simeq V_{\rm I0}/3$ for $\phi_{\rm f} \leq \phi \leq \phi_\star$.

3.2. T-Model Inflation

Although TMI is mostly realized by a gauge-singlet inflaton [5], we here propose an alternative scheme where the inflaton is included in a conjugate pair of Higgs superfields, $\bar{\Phi}$ and Φ with charges $B - L = -1$ and 1 respectively. These fields break the GUT symmetry $G_{\rm GUT} = G_{\rm SM} \times U(1)_{B-L}$ down to the *Standard Model* (SM) gauge group $G_{\rm SM}$ through their *vacuum expectation value* (v.e.vs). We below outline the SUGRA setting in Sec. 3.2.1 and its inflationary outcome in Sec. 3.2.2.

3.2.1. SUGRA Set-up Consistently with the imposed symmetries we adopt the following W

$$W = S\left(\frac{1}{2}\lambda_2\bar{\Phi}\Phi + \lambda_4(\bar{\Phi}\Phi)^2 - \frac{1}{4}M^2\right), \tag{19}$$

where λ_2, λ_4 and M are free parameters. This type of W is well-known from the models of F-term hybrid inflation – for reviews see, e.g., Ref. [12]. The invariance of K under $G_{\rm GUT}$ enforces us to introduce a pole of order two in the $\bar{\Phi} - \Phi$ kinetic terms. One possible option – for another equivalent one see Ref. [2] – is

$$K_{(11)^2} = -\frac{N}{2}\ln\left(1-2|\Phi|^2\right)\left(1-2|\bar{\Phi}|^2\right) \ \text{ or } \ \widetilde{K}_{(11)^2} = -\frac{N}{2}\ln\frac{\left(1-2|\Phi|^2\right)\left(1-2|\bar{\Phi}|^2\right)}{(1-2\bar{\Phi}\Phi)^{1/2}(1-2\bar{\Phi}^*\Phi^*)^{1/2}}, \tag{20}$$

which parameterizes the manifold $\mathcal{M}_{(11)^2} = SU(1,1)/U(1)_\Phi \times SU(1,1)/U(1)_{\bar{\Phi}}$ [2] with scalar curvature $\mathcal{R}_{(11)^2} = -8/N$. From the selected above W and K's, we obtain the models δTM and TM4 – TM8, which employ the following total K's correspondingly

$$\text{(a) } K_{2(11)^2} = K_2 + K_{(11)^2} \ \text{ and } \ \text{(b) } \widetilde{K}_{2(11)^2} = K_2 + \widetilde{K}_{(11)^2}. \tag{21}$$

Within δTM, an elimination of the singular denominator appearing in $V_{\rm I}$ is obtained setting

$$N = 2 \quad \text{and} \quad r_{42} = -\lambda_4/\lambda_2 \simeq 1 + \delta_{42} \quad \text{with} \quad \delta_{42} \sim 0 \quad \text{and} \quad M \ll 1. \tag{22}$$

On the other hand, no singularity exists in $V_{\rm I}$, within TM4 and TM8 and so N, λ_2, λ_4 and M remain free parameters.

3.2.2. Inflationary Configuration Adopting for the relevant fields the parameterization

$$\Phi = \phi e^{i\theta}\cos\theta_\Phi \ \text{ and } \ \bar{\Phi} = \phi e^{i\bar{\theta}}\sin\theta_\Phi \ \text{ with } \ 0 \leq \theta_\Phi \leq \pi/2 \ \text{ and } \ S = (s + i\bar{s})/\sqrt{2}, \tag{23}$$

we can easily verify that a D-flat direction is

$$\langle\theta\rangle_{\rm I} = \langle\bar{\theta}\rangle_{\rm I} = 0,\ \langle\theta_\Phi\rangle_{\rm I} = \pi/4 \ \text{ and } \ \langle S\rangle_{\rm I} = 0, \tag{24}$$

which can be qualified as inflationary path. Regarding the exponential prefactor of $V_{\rm F}$ in Eq. (5) we obtain

$$\langle e^K\rangle_{\rm I} = \begin{cases} f_2^{-N} & \text{for } K = K_{(11)^2}, \\ 1 & \text{for } K = \widetilde{K}_{(11)^2}, \end{cases} \tag{25}$$

Substituting it and Eqs. (9) and (19) into Eq. (8), the inflationary potential $V_{\rm I}$ takes its master form

$$V_{\rm I} = \frac{\lambda^2}{16}\begin{cases} \left(\phi^2 - r_{42}\phi^4 - M_2^2\right)^2/f_2^N & \text{for } \delta\text{TM}, \\ \left(\phi^2 - r_{42}\phi^4 - M_2^2\right)^2 & \text{for TM4}, \\ \left(\phi^4 - r_{24}\phi^2 - M_4^2\right)^2 & \text{for TM8}, \end{cases} \tag{26}$$

where $r_{ij} = -\lambda_i/\lambda_j$ with $i,j = 1,2$ and λ and M_i are identified as follows

$$\lambda = \begin{cases} \lambda_2 \ \text{ and } \ M_2 = M/\sqrt{\lambda_2} & \text{for } \delta\text{TM and TM4}, \\ \lambda_4 \ \text{ and } \ M_4 = M/\sqrt{\lambda_4} & \text{for TM8}. \end{cases} \tag{27}$$

From Eq. (26), we infer that the pole in f_2 is presumably present in $V_{\rm I}$ of δTM but it disappears in $V_{\rm I}$ of TM4 and TM8 and so no N dependence in $V_{\rm I}$ arises. The elimination of the pole in the regime of Eq. (22) lets open the realization of δTM, though.

FIELDS	EIGEN-STATES		MASSES SQUARED $K = K_{2(11)^2}$	$K = \widetilde{K}_{2(11)^2}$
2 real scalars	$\widehat{\theta}_+$	$m^2_{\widehat{\theta}+}$	$3H_{\rm I}^2$	
	$\widehat{\theta}_\Phi$	$\widehat{m}^2_{\theta_\Phi}$	$M_{BL}^2 + 6H_{\rm I}^2(1 + 4/N - 2/N\phi^2 - 2\phi^2/N)$	
1 complex scalar	$s, \bar{s}$	$\widehat{m}^2_s$	$6H_{\rm I}^2(1/N_S - 8(1-\phi^2)/N + N\phi^2/2$ $+2(1-2\phi^2) + 8\phi^2/N)$	$6H_{\rm I}^2(1/N_S - 4/N$ $+2/N\phi^2 + 2\phi^2/N)$
1 gauge boson	A_{BL}	M_{BL}^2	$2Ng^2\phi^2/f_2^2$	
4 Weyl spinors	$\widehat{\psi}_\pm$	$\widehat{m}^2_{\psi\pm}$	$12f_2^2H_{\rm I}^2/N^2\phi^2$	
	$\lambda_{BL}, \widehat{\psi}_{\Phi-}$	M_{BL}^2	$2Ng^2\phi^2/f_2^2$	

Table 2. *Mass spectrum for TMI along the inflationary trajectory of Eq. (24).*

The canonically normalized (hatted) fields of the $\bar{\Phi} - \Phi$ system during TMI, are defined as follows

$$\langle K_{\alpha\bar{\beta}}\rangle_{\rm I} \dot{z}^\alpha \dot{z}^{*\bar{\beta}} \simeq \frac{1}{2}\left(\dot{\widehat{\phi}}^2 + \dot{\widehat{\theta}}_+^2 + \dot{\widehat{\theta}}_-^2 + \dot{\widehat{\theta}}_\Phi^2\right) \quad \text{for} \quad \alpha = 2, 3. \tag{28a}$$

where, via Eq. (20), we find

$$\left(\langle K_{\alpha\bar{\beta}}\rangle_{\rm I}\right) = \langle M_{\Phi\bar{\Phi}}\rangle_{\rm I} \;\text{ with }\; \langle M_{\Phi\bar{\Phi}}\rangle_{\rm I} = \kappa\,\mathsf{diag}(1,1) \;\text{ and }\; \kappa = N/f_2^2. \tag{28b}$$

Inserting the expressions above in Eq. (28a) we obtain the hatted fields

$$d\widehat{\phi}/d\phi = J = \sqrt{2N}/f_2, \;\; \widehat{\theta}_\pm = \sqrt{\kappa}\phi\theta_\pm \;\text{ and }\; \widehat{\theta}_\Phi = \sqrt{2\kappa}\phi\left(\theta_\Phi - \pi/4\right), \tag{29}$$

where $\theta_\pm = \left(\bar{\theta} \pm \theta\right)/\sqrt{2}$. From the first equation above we conclude that Eq. (2) for $p = 2$ is reproduced for $N_2 = 2N$.

We can also verify that the direction of Eq. (24) is stable w.r.t the fluctuations of the non-inflaton fields. Approximate, quite precise though, expressions for $\phi = \phi_\star \sim 1$ are arranged in Table 2. We confine ourselves to the limits $\delta_{42} = M_2 = 0$ for δTM, $r_{42} = M_2 = 0$ for TM4 and $r_{24} = M_4 = 0$ for TM8. As in the case of the spectrum in Table 1, $N_S < 6$ plays a crucial role in retaining positive and heavy enough $\widehat{m}_s^2$. Here, however, we also display the masses, M_{BL}, of the gauge boson A_{BL} (which signals the fact that $U(1)_{B-L}$ is broken during TMI) and the masses of the corresponding fermions. It is also evident that A_{BL} becomes massive absorbing the massless Goldstone boson associated with $\widehat{\theta}_-$.

4. Inflationary Observables

We here constrain the parameters of our EMI and TMI – in Sec. 4.2 and 4.3 respectively – taking into account a number of observational and theoretical requirements described in Sec. 4.1.

4.1. Constraints

We impose to ETI the following observational and theoretical requirements:

(a) The number of e-foldings $N_\star$ that the scale $k_\star = 0.05/\mathrm{Mpc}$ experiences during ETI must be enough for the resolution of the problems of standard Big Bang, i.e., [13]

$$N_\star = \int_{\widehat{\phi}_\mathrm{f}}^{\widehat{\phi}_*} d\widehat{\phi}\frac{V_\mathrm{I}}{V_{\mathrm{I},\widehat{\phi}}} \simeq 61.3 + \frac{1-3w_\mathrm{rh}}{12(1+w_\mathrm{rh})}\ln\frac{\pi^2 g_\mathrm{rh}T_\mathrm{rh}^4}{30V_\mathrm{I}(\phi_\mathrm{f})} + \frac{1}{4}\ln\frac{V_\mathrm{I}(\phi_\star)^2}{g_\mathrm{rh}^{1/3}V_\mathrm{I}(\phi_\mathrm{f})}, \tag{30}$$

where $\widehat{\phi}_*$ is the value of $\widehat{\phi}$ when $k_\star$ crosses the inflationary horizon whereas $\widehat{\phi}_\mathrm{f}$ is the value of $\widehat{\phi}$ at the end of ETI, which can be found, in the slow-roll approximation, from the condition

$$\max\{\epsilon(\phi_\mathrm{f}), |\eta(\phi_\mathrm{f})|\} = 1, \quad \text{where } \epsilon = \frac{1}{2}\left(\frac{V_{\mathrm{I},\widehat{\phi}}}{V_\mathrm{I}}\right)^2 \quad \text{and} \quad \eta = \frac{V_{\mathrm{I},\widehat{\phi}\widehat{\phi}}}{V_\mathrm{I}}. \tag{31a}$$

Also we assume that ETI is followed in turn by an oscillatory phase with mean equation-of-state parameter w_rh, radiation and matter domination. We determine it applying the formula [3]

$$w_\mathrm{rh} = 2\frac{\int_{\phi_\mathrm{mn}}^{\phi_\mathrm{mx}} d\phi J(1-V_\mathrm{I}/V_\mathrm{I}(\phi_\mathrm{mx}))^{1/2}}{\int_{\phi_\mathrm{mn}}^{\phi_\mathrm{mx}} d\phi J(1-V_\mathrm{I}/V_\mathrm{I}(\phi_\mathrm{mx}))^{-1/2}} - 1, \tag{31b}$$

where $\phi_\mathrm{mn} = \langle\phi\rangle$ is the v.e.v of ϕ after ETI. Motivated by implementations [14] of non-thermal leptogenesis, which may follow ETI, we set $T_\mathrm{rh} \simeq 10^9$ GeV for the reheat temperature. Finally, $g_\mathrm{rh} = 228.75$ is the energy-density effective number of degrees of freedom include corresponding to the *Minimal SUSY SM* (MSSM) spectrum.

(b) The amplitude A_s of the power spectrum of the curvature perturbations generated by ϕ at $k_\star$ has to be consistent with data [13], i.e.,

$$A_\mathrm{s} = V_\mathrm{I}(\widehat{\phi}_*)^3/12\,\pi^2 V_{\mathrm{I},\widehat{\phi}}(\widehat{\phi}_*)^2 \simeq 2.105\cdot 10^{-9}\,. \tag{32}$$

(c) The remaining inflationary observables (n_s, its running α_s and r) have to be consistent with the latest *Planck release 4* (PR4), *Baryon Acoustic Oscillations* (BAO), CMB-lensing and BICEP/*Keck* (BK18) data [15, 16], i.e.,

$$\text{(i) } n_\mathrm{s} = 0.965 \pm 0.009 \quad \text{and} \quad \text{(ii) } r \leq 0.032, \tag{33}$$

at 95% *confidence level* (c.l.) – pertaining to the ΛCDM+r framework with $|\alpha_\mathrm{s}| \ll 0.01$. These observables are estimated through the relations

$$\text{(i) } n_\mathrm{s} = 1 - 6\widehat{\epsilon}_\star + 2\widehat{\eta}_\star, \ \text{(ii) } \alpha_\mathrm{s} = \frac{2}{3}\left(4\widehat{\eta}^2 - (n_\mathrm{s}-1)^2\right) - 2\widehat{\xi}_\star \ \text{ and } \ \text{(iii) } r = 16\widehat{\epsilon}_\star\,, \tag{34}$$

with $\xi = V_{\mathrm{I},\widehat{\phi}}V_{\mathrm{I},\widehat{\phi}\widehat{\phi}\widehat{\phi}}/V_\mathrm{I}^2$ – the variables with subscript $\star$ are evaluated at $\phi = \phi_\star$.

(d) The effective theory describing ETI has to remain valid up to a *Ultra Violet* (UV) cutoff scale $\Lambda_\mathrm{UV} \simeq m_\mathrm{P}$ to ensure the stability of our inflationary solutions, i.e.,

$$\text{(i) } V_\mathrm{I}(\phi_\star)^{1/4} \leq \Lambda_\mathrm{UV} \quad \text{and} \quad \text{(ii) } \phi_\star \leq \Lambda_\mathrm{UV}. \tag{35}$$

Finally, we quantify the tuning needed for the attaintment of efficient ETI defining

$$\Delta_\star = 1 - \phi_\star. \tag{36}$$

The naturalness of the achievement of ETI increases with $\Delta_\star$.

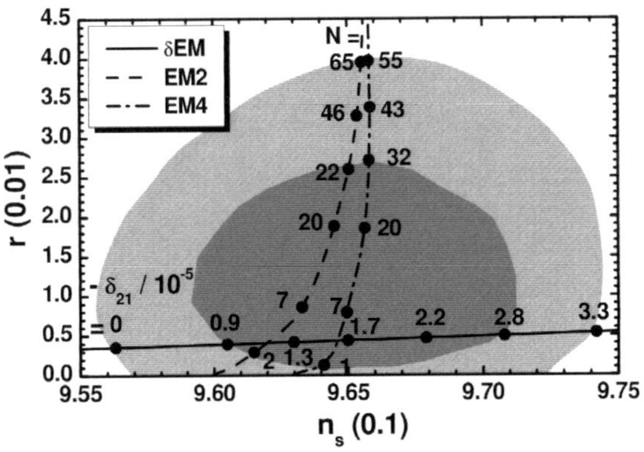

Model:	δEM	EM2	EM4
δ_{21} / r_{21} / r_{12}	$-1.7\cdot10^{-5}$	0.001	0.001
N	2	10	10
$\phi_\star/0.1$	9.9	9.53	9.84
$\Delta_\star(\%)$	1	4.7	2
$\phi_{\rm f}/0.1$	6.66	3.7	5.6
$w_{\rm rh}$	-0.24	-0.08	0.26
$N_\star$	44.4	51.5	55.5
$\lambda/10^{-5}$	1.2	2.1	1.9
$n_{\rm s}/0.1$	9.65	9.64	9.65
$-\alpha_{\rm s}/10^{-4}$	11.4	6.7	6.2
$r/10^{-2}$	0.44	1.3	1.1

Figure 1. *Allowed curves in the $n_{\rm s}-r$ plane for (i) δEM, $M_1=0.01$ and various δ_{21}'s indicated on the solid line or (ii) EM2, $M_1=0.01$ and $r_{21}=0.001$ or EM4, $M_2=0.01$ and $r_{12}=0.001$ and various N's indicated on the dashed or dot-dashed line respectively. The marginalized joint 68% [95%] c.l. regions [15] from PR4, BK18, BAO and lensing data-sets are depicted by the dark [light] shaded contours. The relevant field values, parameters and observables corresponding to points shown in the plot are listed in the Table.*

4.2. E-Model Inflation

After imposing Eqs. (30) and (32) the free parameters of

$$\delta\text{EM},\ \text{EM2},\ \text{EM4}\quad\text{are}\quad(\delta_{21},M_1),\ (N,r_{21},M_1)\ \text{and}\ (N,r_{12},M_2),$$

respectively. Recall that we use $N=2$ exclusively for δEM. Fixing $M_1=0.001$ for δEM, $M_1=0.01$ and $r_{21}=0.001$ for EM2 and $M_2=0.01$ and $r_{12}=0.001$ for EM4, we obtain the curves plotted and compared to the observational data in Fig. 1. We observe that:

(a) For δEM the resulting $n_{\rm s}$ and r increase with $|\delta_{21}|$ – see solid line in Fig. 1. From the considered data we collect the results

$$0\lesssim\delta_{21}/10^{-5}\lesssim3.3,\quad3.5\lesssim r/10^{-3}\lesssim5.3\ \text{and}\ 9\cdot10^{-3}\lesssim\Delta_\star\lesssim0.01. \tag{37}$$

In all cases we obtain $N_\star\simeq44$ consistently with Eq. (30) and the resulting $w_{\rm rh}\simeq-0.237$ from Eq. (31b). Fixing $n_{\rm s}=0.965$, we find $\delta_{21}=-1.7\cdot10^{-5}$ and $r=0.0044$ – see the leftmost column of the Table in Fig. 1.

(b) For EM2 and EM4, $n_{\rm s}$ and r increase with N and $\Delta_\star$ which increases w.r.t its value in δEM. Namely, for EM2 – see dashed line in Fig. 1 – we obtain

$$0.96\lesssim n_{\rm s}\lesssim0.9654,\quad0.1\lesssim N\lesssim65,\quad0.05\lesssim\Delta_\star/10^{-2}\lesssim16.7\ \text{and}\ 0.0025\lesssim r\lesssim0.039 \tag{38a}$$

with $w_{\rm rh}\simeq-0.05$ and $N_\star\simeq50$. On the other hand, for EM4 – see dot-dashed line in Fig. 1 – we obtain

$$0.963\lesssim n_{\rm s}\lesssim0.965,\quad0.1\lesssim N\lesssim55,\quad0.23\lesssim\Delta_\star/10^{-2}\lesssim8.5\ \text{and}\ 0.0001\lesssim r\lesssim0.04 \tag{38b}$$

with $w_{\rm rh}\simeq(0.25-0.39)$ and $N_\star\simeq54-56$. In both equations above the lower bound on N is just artificial. For $N=10$, specific values of parameters and observables are arranged in the rightmost columns of the Table in Fig. 1.

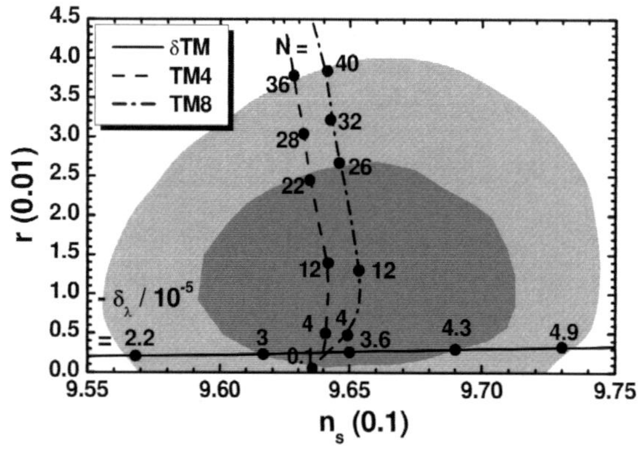

Model:	δTM	TM4	TM8
δ_{42} / r_{42} / r_{24}	$-3.6 \cdot 10^{-5}$	0.01	10^{-6}
N	2	12	12
$\phi_\star/0.1$	9.9555	9.75	9.877
$\Delta_\star(\%)$	0.445	2.5	1.23
$\phi_{\rm f}/0.1$	5.9	3.9	6.5
$w_{\rm rh}$	0.33	0.266	0.58
$N_\star$	55.2	56.4	58
$\lambda/10^{-5}$	3.6	8.6	8.5
$n_{\rm s}/0.1$	9.65	9.64	9.65
$-\alpha_{\rm s}/10^{-4}$	6.6	6.4	5.98
$r/10^{-2}$	0.26	1.4	1.3

Figure 2. *Allowed curves in the $n_{\rm s} - r$ plane fixing $M_{BL} = M_{\rm GUT}$ for (i) δTM and various δ_{42}'s indicated on the solid line or (ii) TM4 and $r_{42} = 0.01$ or TM8 and $r_{24} = 10^{-6}$ and various N's indicated on the dashed and dot-dashed line respectively. The shaded corridors are identified as in Fig. 1. The relevant field values, parameters and observables corresponding to points shown in the plot are listed in the Table.*

4.3. T-Model Inflation

After enforcing Eqs. (30) and (32) – which yield λ together with $\phi_\star$ – the free parameters of the models

$$\delta\text{TM},\ \text{TM4},\ \text{TM8}\ \text{ are }\ (\delta_{42}, M_2),\ (N, r_{42}, M_2)\ \text{ and }\ (N, r_{24}, M_4),$$

respectively. Recall that we use $N = 2$ exclusively for δTM. Also, we determine M_2 and M_4 demanding that the GUT scale within MSSM, $M_{\rm GUT} \simeq 2/2.433 \times 10^{-2}$, is identified with the value of M_{BL} – see Table 2 – at the vacuum value of ϕ, $\langle\phi\rangle$. We approximately obtain M_2 and $M_4 \leq 0.001$. By varying the remaining parameters for each model we obtain the allowed curves in the $n_{\rm s} - r$ plane– see Fig. 2. A comparison with the observational data is also displayed there. We observe that:

(a) For δTM – see the solid line in Fig. 2 – we obtain results similar to those obtained for δEM in Sec. 4.2. Namely, the resulting $n_{\rm s}$ and r increase with $|\delta_{42}|$ with $n_{\rm s}$ covering the whole allowed range in Eq. (33). From the considered data we collect the results

$$2 \lesssim -\delta_{42}/10^{-5} \lesssim 5.5,\quad 2 \lesssim r/10^{-3} \lesssim 3.6\quad \text{and}\quad 4 \lesssim \Delta_\star/10^{-3} \lesssim 4.75. \tag{39}$$

Also, Eq. (30) yields $N_\star \simeq (54.8 - 55.7)$ and Eq. (31b) results to $w_{\rm rh} \simeq 0.3$. Fixing $n_{\rm s} = 0.965$ we find $\delta_{42} = -3.6 \cdot 10^{-5}$ and $r = 0.0026$ – see the leftmost column of the Table in Fig. 2.

(b) For TM4 and TM8, $n_{\rm s}$ and r increase with N and $\Delta_\star$ which is larger than that obtained in δTM. More specifically, for TM4 – see dashed line in Fig. 2 – we obtain

$$0.963 \lesssim n_{\rm s} \lesssim 0.964,\quad 0.1 \lesssim N \lesssim 36,\quad 0.09 \lesssim \Delta_\star/10^{-2} \lesssim 7.6\quad \text{and}\quad 0.0005 \lesssim r \lesssim 0.039, \tag{40a}$$

with $w_{\rm rh} \simeq 0.3$ and $N_\star \simeq 56$. On the other hand, for TM8 – see dot-dashed line in Fig. 2 – we obtain

$$0.963 \lesssim n_{\rm s} \lesssim 0.965,\quad 0.1 \lesssim N \lesssim 40,\quad 0.45 \lesssim \Delta_\star/10^{-2} \lesssim 3.8\quad \text{and}\quad 0.0001 \lesssim r \lesssim 0.039, \tag{40b}$$

with $w_{\rm rh} \simeq (0.25 - 0.6)$ and $N_\star \simeq (54.6 - 60)$. For $N = 12$, specific values of parameters and observables are arranged in the rightmost columns of the Table in Fig. 2.

5. Conclusions

We reviewed the implementation of ETI in the context of SUGRA. We employed as inflaton a gauge singlet or non-singlet superfield for EMI or TMI respectively. The models are relied on the superpotentials in Eqs. (10) and (19) which respect an R symmetry and include, besides inflaton, an inflaton-accompanying field which facilitates the establishment of ETI. In each model we singled out three subclasses of models (δEM, EM2 and EM4) and (δTM, TM4 and TM8). The models δEM and δTM are based on the Kähler potentials in Eq. (12a) and (21a) whereas (EM2, EM4) and (TM4, TM8) in those shown in Eq. (12b) and (21b). Within δEM and δTM any observationally acceptable $n_{\rm s}$ is attainable by tuning δ_{21} and δ_{42} respectively to values of the order 10^{-5}, whereas r is kept at the level of 10^{-3} – see Eqs. (37) and (39). On the other hand, EM2, EM4, TM4 and TM8 avoid any tuning, larger r's are achievable as N increases beyond 2, while $n_{\rm s}$ lies close to its central observational value – see Eqs. (38a) and (38b) for EMI and Eqs. (40a) and (40b) for TMI.

Acknowledgments

Work supported by the Hellenic Foundation for Research and Innovation (H.F.R.I.) under the "First Call for H.F.R.I. Research Projects to support Faculty members and Researchers and the procurement of high-cost research equipment grant" (Project Number: 2251).

References

[1] B.J. Broy, M. Galante, D. Roest and A. Westphal, *J. High Energy Phys.* **12**, 149 (2015) [arXiv:1507.02277]; T. Terada, *Phys. Lett. B* **760**, 674 (2016) [arXiv:1602.07867].
[2] C. Pallis, *J. Cosmol. Astropart. Phys.* **05**, 043 (2021) [arXiv:2103.05534]; C. Pallis, *J. Phys. Conf. Ser.* **2105**, no. 12, 12 (2021) [arXiv:2109.06618].
[3] C. Pallis, *Eur. Phys. J. C* **82**, no. 5, 444 (2022) [arXiv:2204.01047].
[4] J. Ellis, D.V. Nanopoulos and K.A. Olive, *Phys. Rev. Lett.* **111**, 111301 (2013) – Erratum: *Phys. Rev. Lett.* **111**, no. 12, 129902 (2013) [arXiv:1305.1247].
[5] R. Kallosh, A. Linde, and D. Roest, *J. High Energy Phys.* **11**, 198 (2013) [arXiv:1311.0472].
[6] R. Kallosh and A. Linde, *J. Cosmol. Astropart. Phys.* **12**, no. 12, 008 (2021) [arXiv:2110.10902].
[7] J. Ellis, M.A. G. Garcia, D.V. Nanopoulos, K.A. Olive and S. Verner, *Phys. Rev. D* **105**, no. 4, 043504 (2022) [arXiv:2112.04466].
[8] R. Kallosh and A. Linde, *J. Cosmol. Astropart. Phys.* **07**, 002 (2013) [arXiv:1306.5220].
[9] J.J.M. Carrasco, R. Kallosh, A. Linde and D. Roest, *Phys. Rev. D* **92**, no. 4, 041301 (2015) [arXiv:1504.05557].
[10] R. Kallosh, A. Linde and T. Rube, *Phys. Rev. D* **83**, 043507 (2011) [arXiv:1011.5945].
[11] C. Pallis and N. Toumbas, *J. Cosmol. Astropart. Phys.* **05**, no. 05, 015 (2016) [arXiv:1512.05657]; C. Pallis and N. Toumbas, *Adv. High Energy Phys.* **2017**, 6759267 (2017) [arXiv:1612.09202].
[12] C. Pallis, arXiv:0710.3074; R. Armillis and C. Pallis, arXiv:1211.4011.
[13] N. Aghanim *et al.* [*Planck* Collaboration], *Astron. Astrophys.* **641**, A6 (2020) – Erratum: *Astron. Astrophys.* **652**, C4 (2021) [arXiv:1807.06209].
[14] C. Pallis and N. Toumbas, *J. Cosmol. Astropart. Phys.* **12**, 002 (2011) [arXiv:1108.1771]; C. Pallis, *Universe* **4**, no. 1, 13 (2018) [arXiv:1710.05759].
[15] M. Tristram *et al.*, *Phys. Rev. Lett.* **127**, 151301 (2021) [arXiv:2112.07961].
[16] Y. Akrami *et al.* [*Planck* Collaboration], *Astron. Astrophys.* **641**, A10 (2020) [arXiv:1807.06211].

Journal of Physics: Conference Series **2375** (2022) 012013 doi:10.1088/1742-6596/2375/1/012013

Sterile neutrino from D-brane models

Christos Kokorelis

College of Data Science and Engineering, American University of Malta, Triq Dom Mintoff, Bormla, BML 1013, Malta

E-mail: christos.kokorelis@aum.edu.mt

Abstract.

We discuss the appearance of sterile neutrino in D-brane models. In the simplest case, the sterile neutrino (SN) appears from a 5-stack intersecting D6-brane model with a gauged baryon number. SM is mixed with a single generation of left handed active neutrinos and a right handed neutrino.

1. Introduction

The Standard model (SM) of electroweak interactions successfully predicts the Higgs masses and the masses of electroweak gauge bosons, W+, W-, Z. However, there are intrinsic weaknesses; the SM does not describe gravitational interactions or solve the strong CP-problem, the cosmological constant problem, the flavour puzzle. It also conserves B-L symmetry and thus is allows the baryon number non-conserving interactions [1]

$$\begin{array}{ll} (\bar{d}_R^C u_R)(\bar{q}_L^C l_L), & (\bar{q}_L^C q_L)(\bar{u}_R^C l_R) \\ (\bar{q}_L^C q_L)(\bar{q}_L^C l_L), & (\bar{q}_{iL}^C q_{jL})(\bar{q}_{kL}^C l_{lL})(\vec{\tau}\epsilon)_{ij}(\vec{\tau}\epsilon)_{kl} \\ (\bar{d}_R^C u_R)(\bar{u}_R^C l_R), & (\bar{u}_R^C u_R)(\bar{d}_R^C l_R) \end{array} .$$

As these interactions are non-renormalizable, and are part of a "higher" ultaviolet theory, they are effectively suppressed by the inverse mass squared of the "higher" effective theory M_s. For the proton decay mode $p \to e+\pi^0$ the proton lifetime is estimated at $\tau_p > 10^{35}$ yrs (greater than the age of the universe) as long as $M_s \approx 10^{16}$ GeV. However, in this way the gauge hierarchy problem is introduced, namely an unwanted particle desert between the electroweak scale and the new M_s scale is generated. Nevertheless, SM baryon number is violated by sphaleron processes [2] (see also [3]) non-perturbatively. It is well known that dynamics of galaxies and galaxy clusters cannot be explained by the Newtonian potential created by just the visible SM matter. Also cosmological data analysis reveals that the cosmic large scale structure began to develop before decoupling of photons due to the recombination of hydrogen in the early Universe, namely before ordinary matter start clustering. This evidence points at the existence of new matter, the so called Dark matter (DM), weakly interacting with SM matter, therefore providing a contribution to the total energy density of the Universe at approximately 27%. As a result, a background of relic neutrinos was created before primordial nucleosynthesis; nevertheless as the neutrinos cannot constitute 100% of SM DM, other particles may be introduced, one example is the so called sterile neutrinos (see [4] for a review), SM singlets that could account for the missing DM, especially if their masses are at the KeV region [5] and also generate the baryon asymmetry of the universe using leptogenesis [6].

On the other hand recent experiments suggest the existence of "sterile" neutrino species even though no sterile neutrino has been actually observed in experiments. Earlier results on the existence of a fourth sterile neutrino family were provided by the LSND experiment [7] and non-completely confirmed by MiniBooNE experiment [8]. Other anomalies found in reactor neutrino data could have explained the data by invoking the existence of an eV-mass sterile neutrino, however uncertainties still exist on the reactor energy spectra [9]. Additional anomalies were also found by the experiments Gallium, GALLEX , SAGE [10] and BEST [11]. As result, global fits using the existence of extra sterile neutrino families easily accommodate some datasets but are in tension with other ones [12].

Also, experimental data from solar, atmospheric, reactor and accelerator experiments [13] (namely, SNO, Super-Kamiokande, Icecube, KamLand, DayaBay, RENO, Double Choz, T2K, Nova, Minos, Opera) exhibit neutrino oscillations, where the weak eigenstates mix with the mass eigenstates as described by PMNS unitary mixing matrix formalism [14]. The unitary matrix in this respect are described by three mixing angles θ_{12}, θ_{13}, θ_{23} and a CP violating phase δ_{CP}. Nevertheless, there are also anomalies observed as follows :
i) In short baseline $\bar{\nu}_e$ beams of LSND [7] and MiniBooNE [8] experiments the non-canonical appearance of $\bar{\nu}_e$ is observed,
ii) In gallium experiments a decrement of ν_e predicted rates from radioactive sources is seen.

The new "observed" data could be also described by a forth sterile neutrino (named sterile neutrino), with mass m_4, that does not couple through weak interaction. Experimental results constrain the sterile neutrino mass m_4 squared differences suggesting

$$10^{-4}\ \frac{eV^2}{c^4}\ < \triangle m_{41}^2 = m_4^2 - m_1^2 < 3\times 10^{-3}\ \frac{eV^2}{c^4}\ ; \qquad (1)$$

the lower limit comes from Minos experiment[15], where the assumed 3+1 model, suggests $\triangle m_{32}^2 \approx \triangle m_{31}^2$; and $\triangle m_{41}^2 >> \triangle m_{31}^2$, such that $\triangle m_{41}^2 \approx \triangle m_{42}^2 \approx \triangle m_{43}^2$);
the upper limit from T2K Superkamiokande [16] experiments.

The DANSS [18, 19], NEOS [20, 21], STEREO [22, 23], PROSPECT [24] oscillation experiments suggest $sin^2\theta_{41} = 2.4$ eV^2 at > 90 % CL.

Let us note that it is generally believed that any global symmetries, including necessarilly the baryon number (B) should be gauged. However, unified models based on supersymmetric (or not) field theory and/or supergravity models or even string theory models in various formulations do not incorparate B as a gauged symmetry. The only string theory models [1] gauging B are the so called Madrid group string models, using four -[28], five -[29] and six-[30] D6-brane [2] stacks of branes intersecting (IB) at angles.

SM has no explanation for the origin of neutrino masses, as there no neutrino mass term allowed by the SM gauge symmetries. A mass term for the neutrinos is introduced, which uses right handed neutrinos ν_R through the term $Y_D l_L \nu_R H$, where $l = (\nu, e)_L^T$, the lepton doublet and H the Higgs mass. The Yukawa coupling Y_D has to be of order $\sim 10^{-11}$ to be consistent with neutrino masses of 1 eV, which is quite unnatural to be achieved. On the contrary, in IB models Y_D could be naturally suppressed, as e.g. $Y_D \sim e^{-A}$, where A is the worldsheet area, $0 \le A \le 1$, between the branes ν_L, ν_R and H generating the neutrino Yukawa coupling.

We found that string theory intersecting D-brane [31, 32, 33] models, examining one particular representative, the 5-stack models of [29], possess sterile neutrino candidates.

[1] Proton decay in SU(5) IB, contributing to $p \to e + \pi^0$, $10^2\bar{1}0^2$ was calculated in [25]. On [26] the proton decay mode $p \to \bar{\nu}_L \pi^+$ was exhibited for SU(5) $(10 \cdot 10 \cdot \bar{5} \cdot \bar{5})$ and flipped SU(5) $(10 \cdot 5 \cdot \bar{10} \cdot \bar{5}$) GUTS from intersecting leftbranes. In [27] the proton decay mode, suggested in [26] for SU(5) GUTS was calculated.

[2] D6-branes are 7-dimensional hypersurfaces intersecting at angles, where open strings stretched between any two of them represent the four dimensional fermions.

2. The five stack D6-brane model

Intersecting brane constructions [31, 32, 33, 34] localize chiral fermions as open strings stretching between brane intersecting at angles and gauge bosons living on D-branes. D-branes give rise to a U(1) and the generated U(N) gauge group is associated to N D-brane stacks. By considering N_a stacks of D-brane configurations with $N_a, a = 1, \cdots, N$, parallel branes, a gauge group $U(N_1) \times U(N_2) \times \cdots \times U(N_a)$ appears. Each $U(N_i)$ factor will give rise to a $SU(N_i)$, charged under the associated $U(1_i)$ gauge group that appears in the decomposition $SU(N_a) \times U(1)_a$. The massless spectrum of the model is seen in table (1) .

Matter Fields		Intersection	Q_a	Q_b	Q_c	Q_d	Q_e	Y
Q_L	$(3,2)$	$I_{ab}=1$	1	-1	0	0	0	1/6
q_L	$2(3,2)$	$I_{ab^*}=2$	1	1	0	0	0	1/6
U_R	$3(\bar{3},1)$	$I_{ac}=-3$	-1	0	1	0	0	$-2/3$
D_R	$3(\bar{3},1)$	$I_{ac^*}=-3$	-1	0	-1	0	0	1/3
L	$2(1,2)$	$I_{bd}=-2$	0	-1	0	1	0	$-1/2$
l_L	$(1,2)$	$I_{be}=-1$	0	-1	0	0	1	$-1/2$
N_R	$2(1,1)$	$I_{cd}=2$	0	0	1	-1	0	0
E_R	$2(1,1)$	$I_{cd^*}=-2$	0	0	-1	-1	0	1
ν_R	$(1,1)$	$I_{ce}=1$	0	0	1	0	-1	0
e_R	$(1,1)$	$I_{ce^*}=-1$	0	0	-1	0	-1	1

Table 1. Chiral fermionic spectrum of the five stack string scale $SU(3)_C \otimes SU(2)_L \otimes U(1)_a \otimes U(1)_b \otimes U(1)_c \otimes U(1)_d \otimes U(1)_e$ intersecting D6-brane model. At low energies only the SM gauge group $SU(3) \otimes SU(2)_L \otimes U(1)_Y$ survives.

The initial gauge group of the model is $U(3)_c \times U(2)_b \times U(1)_c \times U(1)_d \times U(1)_e$ or $SU(3)_c \times SU(2)_w \times U(1)_a \times U(1)_b \times U(1)_c \times U(1)_d \times U(1)_e$ at the string scale. The model accommodates the global symmetries of the Standard model (SM) , namely Baryon B and Lepton number L, to local gauge symmetries. There are various gauged low energy symmetries in the models. They are defined in terms of the U(1) symmetries Q_a, Q_b, Q_c, Q_d, Q_e, where the baryon number B and lepton number L, respectively are equal to

$$Q_a = 3B,\ L = Q_d + Q_e,\ Q_a - 3Q_d - 3Q_e = 3(B-L),\ Q_c = 2I_{3R} \tag{2}$$

and I_{3R} being the third component of weak isospin and $3(B-L)$ and Q_c are free of triangle anomalies. The $U(1)_b$ symmetry plays the role of a Peccei-Quinn symmetry, having mixed SU(3) anomalies. The hypercharge is defined as $Y = \frac{1}{6}Q_a - \frac{1}{2}Q_c - \frac{1}{2}Q_d - \frac{1}{2}Q_e$.

Our SM can be embedded in a bottom-up approach in string compactification of IIA theory on a six dimensional torus equipped with an orientifolded symmetry which converts, strings with D9-branes with fluxes compactified on a six-dimensional orientifolded torus T6 (where internal background gauge fluxes on the branes are turned on) by the use of a T-duality transformation on the x4, x5, x6, directions, into D6-branes intersecting at angles. In detail, we assume that the D6$_a$-branes are wrapping 1-cycles (n_a^i, m_a^i) along each of the ith-T^2 torus of the factorized T^6 torus, namely $T^6 = T^2 \times T^2 \times T^2$. That means that we allow our torus to wrap factorized 3-cycles, that can unwrap into products [3] of three 1-cycles Π_a, one for each T^2. We define the

[3] We define the homology of the 3-cycle as $\Pi_a = \prod_{i=1}^{3}(n_a^i[a_i] + m_a^i[b_i])$. Because of the orientifold ΩR symmetry, where Ω is the worldvolume parity and R is the reflection on the T-dualized coordinates, $T(\Omega R)T^{-1} = \Omega R$, each D6$_a$-brane 3-cycle, must have its ΩR orientifold image $\Pi_{a^\star} = \prod_{i=1}^{3}(n_a^i[a_i] - m_a^i[b_i])$.

number of chiral fermions, i.e. the number of electrons, that are located at intersections of the branes a, b to be equal to the homology product of 3-cycles as

$$I_{ab} = [\Pi_a] \cdot [\Pi_b] = \prod_{i=1}^{3} (n_a^i m_b^i - m_a^i n_b^i) \tag{3}$$

and transforming in the bifundamental representation $(N_a, \bar{N}_b)$; for a left handed fermion with $I_{ab} > 0$. We also define the intersection number that determines the number of chiral fermions at the intersection of the a-brane with the orientifold image of the b-brane, namely

$$I_{ab^\star} = [\Pi_a] \cdot [\Pi_{b^\star}] = -\prod_{i=1}^{3} (n_a^i [a_i] + m_a^i [b_i]) \tag{4}$$

The fermions at the ab* intersection transform at the representation (N_a, N_b) for a left handed fermion with $I_{ab*} > 0$. Vacua subject to the intersection constraints (3), are also subject to additional constraints coming from RR tadpole cancellation [29], [32]. That demands cancellation of D6-branes charges, wrapping on three cycles with homology $[\Pi_a]$ and O6-plane 7-form charges wrapping on 3-cycles with homology $[\Pi_{O_6}]$. The model predicts the existence of one (1) or two (2) sneutrinos which can be used to break the extra, beyond the SM gauge symmetry associated with the extra neutral Z$'$ gauge boson associated with the gauge symmetry Q_N using the sν_R [29]. In [35] this mechanism was also used to break the U(1) B-L gauged symmetry in a GUT gauge theory model. We note that gauging B-L cannot avoid generating baryon number violating non-renormalizable interactions as in the SM. The model [29] also obeys $b \to s l^+ l^-$ anomalies and furthermore its stringy Z$'$ gauge boson has nonnegligible couplings to the first two quark generations and has a mass in the range [3.5, 5.5] TeV, so it is possible to be discovered directly during the LHC runs via Drell-Yan production in the di-electron or di-muon decay channels [36].

Active neutrinos could get a mass from Yukawa couplings or from dimension 6 operators [29]. A mechanism for generating small active neutrino masses $\approx$ 0.1-10 eV in consistency with LSND experiments [29] is generated from dimension 6 operators , $\alpha'(LN_R) < Q_L U_R >$, $\alpha'(l\nu_R)(< q_L U_R >)$ which are associated with the breaking of $U(1)_b$ PQ like symmetry through chiral symmetry breaking related to the existence [28] of the u-quark chiral condensate $< Q_L U_R > = < q_L U_R > \approx 240\ (\mathrm{MeV})^3$.

3. The sterile neutrino in D-brane models

In IB constructions the right handed neutrino ν_R is part of the D-brane SM matter spectrum and thus IB constructions is a promising framework, that can be used to obtain predictions from string theory [36, 37] and also naturally accommodate models with ν_R's and a stable proton[29].

In a field theory model a sterile neutrino mass is generated

$$M_{gauge} = \begin{pmatrix} 0 & m_1 & 0 \\ m_1 & 0 & m_2 \\ 0 & m_2 & m_3 \end{pmatrix}, \tag{5}$$

by the mass terms (see e.g. [38]),

$$m_1 \nu_R \nu_L + m_2 \nu_R N + m_3 N N \tag{6}$$

where M_{GUT} is the gauge unification scale and H, $\bar{K}$ Higgses and ν_R a right handed neutrino. The presence of Majorana mass term for the sterile state N is evident. The neutrino mass generation of (5) is called the inverse see-saw mechanism [39, 40]. See also [38]. In D-brane models, we obtain [41] that Dirac-type interactions, mix a single generation of left handed neutrino with a single generation of a right handed neutrino and the sterile neutrino N; no Majorana term for the sterile neutrino mass is obtained as in (5))

$$\mathcal{L} = m_D \nu_L \nu_R + m_N \nu_R N + m_S \nu_L N \tag{7}$$

generating an effective 3 x 3 mass matrix

$$M = \begin{pmatrix} 0 & m_D & m_S \\ m_D & 0 & m_N \\ 0 & m_N & 0 \end{pmatrix}, \tag{8}$$

where the values of m_N and m_S are obtained by non-renormalizable string interactions [41]. Depending on the values of m_N, m_D, m_S the lowest state can be either the left neutrino or the sterile neutrino. Our mechanism is a variant of the inverse see-saw mehanism [39, 40]. The present models have the clear advantage that they localize the baryon number as a gauged symmetry, thus proton decay interactions are absent. We note that recently sterile neutrino mass studies in other formulations of string theory [42, 43], namely fermionic formulation and F-theory respectively have appeared, however on these constructions baryon number is not a gauge symmetry.

References

[1] S. Weinberg, "Baryon and Lepton Nonconserving Processes", Phys. Rev. Lett. 43 (1979) 1566
[2] G. 't Hooft, Phys. Rev. Lett. 37, 37 (1976)
[3] R.F. Klinkhamer and N.S. Manton, Phys. Rev. D30, 2212 (1984)
[4] A. Boyarsky, M. Drewes, T. Lasserre, S. Mertens, O. Ruchayskiy, Sterile neutrino Dark Matter, Prog. Part. Nucl. Phys. 104 (2019) 1-45, e-Print: 1807.07938 [hep-ph]
[5] S. Dodelson, L. M. Widrow, Sterile-neutrinos as dark matter, Phys.Rev.Lett. 72 (1994) 17–20. doi:10.1103/PhysRevLett.72.17. ;X.-D. Shi, G. M. Fuller, A New dark matter candidate: Nonthermal sterile neutrinos, Phys.Rev.Lett. 82 (1999) 2832–2835. doi:10.1103/PhysRevLett.82.2832 ; A. Dolgov, S. Hansen, Massive sterile neutrinos as warm dark matter, Astropart.Phys. 16 (2002) 339–344. doi:10.1016/S0927-6505(01)00115-3 ; K. Abazajian, G. M. Fuller, W. H. Tucker, Direct detection of warm dark matter in the X-ray, Astrophys. J. 562 (2001) 593–604. doi:10.1086/323867; K. Abazajian, G. M. Fuller, M. Patel, Sterile neutrino hot, warm, and cold dark matter, Phys.Rev. D64 (2001) 023501. doi:10.1103/PhysRevD.64.023501; T. Asaka, S. Blanchet, M. Shaposhnikov, The nuMSM, dark matter and neutrino masses, Phys.Lett. B631 (2005) 151–156. doi:10.1016/j.physletb.2005.09.070
[6] T. Asaka, M. Shaposhnikov, The nuMSM, dark matter and baryon asymmetry of the universe, Phys.Lett. B620 (2005) 17–26. doi:10.1016/j.physletb.2005.06.020. [38] T. Asaka, M. Laine, M. Shaposhnikov, Lightest sterile neutrino abundance within the nuMSM, JHEP 0701 (2007) 091. doi:10.1088/1126-6708/2007/01/091
[7] A. Aguilar et al. (LSND Collaboration), Phys. Rev. D 64, 112007 (2001)
[8] A. A. Aguilar-Arevalo et al. (MiniBooNE Collaboration), Phys. Rev. Lett. 121, 221801 (2018)
[9] J. M. Berryman, P. Coloma, P. Huber, T. Schwetz, and A. Zhou, "Statistical significance of the sterile-neutrino hypothesis in the context of reactor and gallium data," JHEP 02, 055 (2022), arXiv:2111.12530 [hep-ph]
[10] Carlo Giunti and Marco Laveder, "Statistical Significance of the Gallium Anomaly," Phys. Rev. C 83, 065504 (2011), arXiv:1006.3244 [hep-ph]
[11] V. V. Barinov et al., "Results from the Baksan Experiment on Sterile Transitions (BEST)," Phys. Rev. Lett. 128, 232501 (2022), arXiv:2109.11482 [nucl-ex]; V. V. Barinov and D. Gorbunov, BEST impact on sterile neutrino hypothesis, Phys. Rev. D 105, L051703 (2022), arXiv:2109.14654 [hep-ph
[12] M. Dentler, Á. Hernández-Cabezudo, J. Kopp, P. A. N. Machado, M. Maltoni, I. Martinez-Soler, and T. Schwetz, Updated Global Analysis of Neutrino Oscillations in the Presence of eV-Scale Sterile Neutrinos, JHEP 08, 010 (2018), arXiv:1803.10661 [hep-ph]; C. A. Argüelles, I. Esteban, M. Hostert, K. J. Kelly, J. Kopp, P. A. N. Machado, I. Martinez-Soler, and Y. F. Perez-Gonzalez, MicroBooNE and the ν_e Interpretation of the MiniBooNE Low-Energy Excess, Phys. Rev. Lett. 128, 241802 (2022), arXiv:2111.10359 [hep-ph]; C. Giunti, Y. F. Li, C. A. Ternes, O. Tyagi, and Z. Xin, Gallium Anomaly: Critical View from the Global Picture of ν_e and $\bar{\nu}_e$ Disappearance, (2022), arXiv:2209.00916 [hep-ph]

[13]] B. Aharmim et al. (SNO Collaboration), Phys. Rev. C 88, 025501 (2013); K. Abe et al. (Super-Kamiokande Collaboration), Phys. Rev. D 94, 052010 (2016); K. Abe et al. (Super-Kamiokande Collaboration), Phys. Rev. D 97, 072001 (2018); M. G. Aartsen et al. (IceCube Collaboration), Phys. Rev. Lett. 120, 071801 (2018); A. Gando et al. (KamLAND Collaboration), Phys. Rev. D 88, 033001 (2013); D. Adey et al. (Daya Bay Collaboration), Phys. Rev. Lett. 121, 241805 (2018); G. Bak et al. (RENO Collaboration), Phys. Rev. Lett. 121, 201801 (2018); Y. Abe et al. (Double Chooz Collaboration), JHEP 01, 163 (2016); K. Abe et al. (T2K Collaboration), Phys. Rev. Lett. 121, 171802 (2018); M. A. Acero et al. (NOvA Collaboration), Phys. Rev. D 98, 032012 (2018); N. Agafonova et al. (OPERA Collaboration), Phys. Rev. Lett. 120, 211801 (2018)

[14] B. Pontecorvo, Sov. Phys. JETP 6, 429 (1957); B. Pontecorvo, Sov. Phys. JETP 26, 984 (1968);Z. Maki, M. Nakagawa, and S. Sakata, Prog. Theor. Phys. 28, 870 (1962)

[15] P. Adamson et al. (MINOS Collaboration), Search for sterile neutrinos in MINOS and MINOS+ using a two-detector fit, Phys. Rev. Lett. 122, 091803 (2019), arXiv:1710.06488 [hep-ex]

[16] K. Abe et al., Search for light sterile neutrinos with the T2K far detector Super-Kamiokande at a baseline of 295 km, Phys. Rev. D 99, 071103 (2019), arXiv:1902.06529 [hep-ex]

[17] A. Minotti, Status of the search for light sterile neutrinos at short baselines, PoS NuFact2021 (2022) 246

[18] I. Alekseev, V. Belov, V. Brudanin, M. Danilov, V. Egorov, D. Filosofov et al., Search for sterile neutrinos at the danss experiment, Physics Letters B 787 (2018) 56

[19] M. Danilov and N. Skrobova, New results from the danss experiment, arXiv preprint arXiv:2112.13413 (2021)

[20] Y. Ko, B. Kim, J. Kim, B. Han, C. Jang, E.J. Jeon et al., Sterile neutrino search at the neos experiment, Physical review letters 118 (2017) 121802

[21] Y. Ko, B.-Y. Han, C.-H. Jang, E.-J. Jeon, K.-K. Joo, B.-R. Kim et al., Neos experiment, in Journal of Physics: Conference Series, vol. 1216, p. 012004, IOP Publishing, 2019

[22] H. Almazán, P. del Amo Sanchez, L. Bernard, A. Blanchet, A. Bonhomme, C. Buck et al., Sterile neutrino constraints from the stereo experiment with 66 days of reactor-on data, Physical review letters 121 (2018) 161801.

[23] H. Almazán, L. Bernard, A. Blanchet, A. Bonhomme, C. Buck, P. del Amo Sanchez et al., Improved sterile neutrino constraints from the stereo experiment with 179 days of reactor-on data, Physical Review D 102 (2020) 052002

[24] M. Andriamirado, A. Balantekin, H. Band, C. Bass, D. Bergeron, D. Berish et al., Improved short-baseline neutrino oscillation search and energy spectrum measurement with the prospect experiment at hfir, Physical Review D 103 (2021) 032001

[25] I. Klebanov and E. Witten, Proton decay in intersecting D-brane models, Nucl.Phys. B 664 (2003) 3, e-Print: hep-th/0304079 [hep-th]

[26] M. Axenides, E. Floratos and C. Kokorelis, SU(5) Theories from intersecting branes, JHEP 10 (2003) 006, e-Print: hep-th/0307255 [hep-th]

[27] M. Cvetic, R. Richter, Proton decay via dimension-six operators in intersecting D6-brane models, Nucl. Phys. B 762 (2007) 112, e-Print: hep-th/0606001 [hep-th]

[28] L. Ibanez, F. Marchesano and R. Rabadan, Getting just the standard model at intersecting branes, JHEP 11 (2001) 002, e-Print: hep-th/0105155 [hep-th]

[29] C. Kokorelis, New standard model vacua from intersecting branes, JHEP 09 (2002) 029, e-Print: hep-th/0205147 [hep-th]

[30] C. Kokorelis, Exact standard model compactifications from intersecting branes, JHEP 08 (2002) 036, e-Print: hep-th/0206108 [hep-th]

[31] D. Lust, Intersecting Brane Worlds – A Path to the Standard Model ? Class. Quant. Grav. 21 (2004) S1399, arXiv:hep-th/0401156[hep-th]

[32] R. Blumenhagen, B. Kors, D. Lust, S. Stieberger, Four-dimensional String Compactifications with D-Branes, Orientifolds and Fluxes, Phys. Rept. 445 (2007) 1, arXiv: hep-th/0610327 [hep-th]

[33] R. Blumenhagen, M. Cvetic, P. Langacker, G. Shiu, Toward realistic intersecting D-brane models Ann. Rev. Nucl. Part. Sci. 55 (2005) 71, arXiv: hep-th/0502005 [hep-th]

[34] C. Angelantonj and A. Sagnotti, 'Open Strings", Phys. Rept. 371 (2002) 1; Phys.Rept. 376 (2003) 6, 407 (erratum) arXiv: hep-th/0204089 [hep-th]

[35] V. Barger, P.F. Perez and S. Spinner, Minimal gauged U(1)(B-L) model with spontaneous R-parity violation, Phys.Rev.Lett. 102 (2009) 181802 , e-Print: 0812.3661 [hep-ph]

[36] A. Celis, W. Z. Feng and D. Lust, "Stringy explanation of $b \to sl^+l^-$ anomalies," JHEP 1602, 007 (2016), arXiv:1512.02218 [hep-ph]

[37] L. A. Anchordoqui, I. Antoniadis, X. Huang, D. Lust, T. R. Taylor, Muon Discrepancy Within D-brane String Compactifications, Fortsch. Phys. 69 (2021) 8-9, 2100084, e-Print: 2104.06854 [hep-ph]

[38] G. K. Leontaris and Q. Shafi, Light sterile neutrino and low scale left-right symmetry in a D-brane inspired

$SU(4)_C \times SU(2)_L \times SU(2)_R$ model, Phys. Rev. D **99**, no. 3, 035032 (2019), [arXiv:1811.12647 [hep-ph]]
[39] R. N. Mohapatra and J. W. F. Valle, Phys. Rev. D 34 (1986) 1642
[40] M. Malinsky, J. C. Romao and J. W. F. Valle, Phys. Rev. Lett. 95 (2005) 161801 [hep-ph/0506296
[41] I. Antoniadis and C. Kokorelis, Sterile neutrinos from D-brane models, to appear
[42] A. E. Faraggi and M. Guzzi, Z′s and sterile neutrinos from heterotic string models: exploring Z′ mass exclusion limits, Eur. Phys. J. C 82 (2022) 7590, e-Print: 2204.11974 [hep-ph]
[43] V. Basiouris, G. K. Leontaris, Sterile neutrinos, $0\nu\beta\beta$ decay and the W-boson mass anomaly in a Flipped SU(5) from F-theory, e-Print: 2205.00758 [hep-ph]

AUTHOR INDEX

Alexopoulos, T 51
Alexopoulos, Theodoros 31
Amperiadou, D 62
Apostolopoulos, T 51
Bandekas, D V 51
Bignami, A 51
Bita, Daniel 31
Bourlis, G 72
Chang, P 42
Dordevic, Milos 21
Drivas-Koulouris, I. 92
Elmer, P 42
Gazis, E 51
Gazis, N 51
Gkialas, I 72
Gu, Y 42
Karagiannaki, A 51
Karentzos, Efstathios 92
Kitsaki, Chara 31
Kokorelis, Christos 120
Kolitsi, Foteini 31
Kordas, K 62
Kotsopoulos, D 51
Krutelyov, V 42
Kyriakis-Bitzaros, Efstathios 31
Lampoudis, C 62
Leisos, A 62, 72
Manthos, I 72
Mesolongitis, Ioannis 31
Mitsou, Vasiliki A 11
Narayanan, B V Sathia 42
Niendorf, G 42
Nonis, S 72
Pallis, C 110
Papageorgiou, K 72
Paraskevopoulos, Christos 82
Petridou, C 62
Pramatari, K 51
Reid, M 42
Sampsonidis, D 62
Tadel, M 42
Telali, I 51
Trachanas, E 51
Tsiafis, Ch 62
Tsirigotis, A 62, 72
Tzamarias, S 62, 72
Tzanis, Polyneikis 100
Vellidis, Costas 1
Vourliotis, E 42
Wang, B 42
Wittich, P 42
Yagil, A 42
Zachariadou, Katerina 31